# SpringerBriefs in Computer Science

More information about this series at http://www.springer.com/series/10028

Ye Tian · Min Zhao · Xinming Zhang

# Internet Video Data Streaming

## Energy-Saving and Cost-Aware Methods

Springer

Ye Tian
School of Computer Science
  and Technology
University of Science
  and Technology of China
Hefei, Anhui
China

Xinming Zhang
School of Computer Science
  and Technology
University of Science
  and Technology of China
Hefei, Anhui
China

Min Zhao
School of Computer Science
  and Technology
University of Science
  and Technology of China
Hefei, Anhui
China

ISSN 2191-5768          ISSN 2191-5776   (electronic)
SpringerBriefs in Computer Science
ISBN 978-981-10-6522-4          ISBN 978-981-10-6523-1   (eBook)
https://doi.org/10.1007/978-981-10-6523-1

Library of Congress Control Number: 2017954878

Printed on acid-free paper

This Springer imprint is published by Springer Nature
The registered company is Springer Nature Singapore Pte Ltd.
The registered company address is: 152 Beach Road, #21-01/04 Gateway East, Singapore 189721, Singapore

# Preface

Video has become more and more prevalent on the Internet. Studies show that YouTube has already accounted for 19.27 and 13.19% of the Internet traffics in Europe and in North America, respectively, in 2014, and it is estimated that by the year 2020, 82% of the global IP traffic will be video content. Most of Internet videos are distributed through Content Delivery Networks (CDNs). A CDN is a large distributed system consisting hundreds of thousands of dedicated content servers. Unlike peer-to-peer video streaming networks, running a CDN for holding a large-scale video service is expensive, as there are considerable costs on bandwidth usage as well as the energy consumed by the network servers.

In this book, we first review the key issues and design choices of several representative CDNs, for instance, Akamai and Google, and discuss the energy-saving techniques for server clusters, data centers, and CDNs. We then tackle the problem of saving a video streaming CDN's operating expense, including both its energy cost and the traffic cost. From our measurement study on the CDN infrastructure of Youku, which is the largest video service site in China, we find that there exists an inherent conflict between improving a video streaming CDN's energy efficiency for power saving and maintaining the CDN's ISP-friendly server selection policy. To solve this conflict, we propose a cost-aware capacity provisioning algorithm, which dynamically plans the service capacities of a CDN's server clusters in numerous ISPs and optimizes its overall operating cost regarding both the energy consumptions and the cross-ISP traffics. By using the workload derived from real-world measurement and applying actual power and bandwidth price parameters, we show with simulation experiments that our approach can significantly reduce a video streaming CDN's overall operating cost and avoid frequent server switches effectively. Finally, we discuss the directions for the future research.

For understanding the book, readers should have general knowledge on computer networks and algorithms. We hope the book can be helpful for the audience, in particular for the researchers in the Internet and networking area.

Hefei, Anhui, China                                                   Ye Tian
August 2017                                                          Min Zhao
                                                                  Xinming Zhang

# Acknowledgements

We would like to thank our colleagues, who provide valuable insights and advices during the book preparation. We also want to thank the editors and staff at Springer for their assistance. This book is not possible without their enormous help. Finally, we owe our gratitude to our families, who support us throughout this book.

We would like to acknowledge the supports from the National Natural Science Foundation of China (Project Nos. 61672486, 61379130, and 61672485), Key Project of the New Generation Mobile Wireless Broadband Communication Networks from MIIT of China (Project No. 2017ZX03001019-004), and the Anhui Provincial Natural Science Foundation (Project No. 1608085MF126).

Acknowledgements

# Contents

**1 Introduction** .................................................... 1
  1.1  Background ................................................ 1
  1.2  CDN Objectives .......................................... 2
      1.2.1  Meeting Service Level Agreement ................ 2
      1.2.2  Reducing Traffic Expense ........................ 3
      1.2.3  Reducing Energy Expense ........................ 4
  1.3  Challenges in CDN ........................................ 4
      1.3.1  System Architecture ............................. 4
      1.3.2  Server Selection ................................ 5
      1.3.3  Capacity Provisioning .......................... 6
  1.4  Overview of the Book .................................... 6
  References ................................................... 7

**2 Content Delivery Networks and Its Interplay with ISPs** .......... 9
  2.1  CDNs in Real World ...................................... 9
      2.1.1  Akamai ......................................... 9
      2.1.2  Limelight ....................................... 11
      2.1.3  Amazon CloudFront ............................ 12
      2.1.4  Google ......................................... 12
      2.1.5  Bing ........................................... 13
      2.1.6  YouTube ....................................... 14
      2.1.7  Netflix and Hulu ............................... 15
      2.1.8  CDNs in China ................................. 15
  2.2  Interplay Between CDN and ISP .......................... 16
      2.2.1  ISP Prospective ................................ 16
      2.2.2  CDN Prospective ............................... 16
  References ................................................... 17

**3  Energy Management** .................................................... 19
   3.1  Energy Saving for Data Center ................................... 19
        3.1.1  DVFS-Based Energy Saving ................................ 19
        3.1.2  Energy Saving with Virtualization ...................... 20
        3.1.3  Energy Saving with Capacity Right-Sizing ............... 21
   3.2  Energy Saving for CDN .......................................... 22
        3.2.1  Energy-Aware Load Balancing ............................ 22
        3.2.2  Battery-Based Power Saving ............................. 22
   References ......................................................... 23

**4  Cost Measurement for Internet Video Streaming** ..................... 25
   4.1  Measurement Methodology and CDN Architecture ................... 25
   4.2  Server Selection Policy Analysis ............................... 28
        4.2.1  Server Selection Characteristics ....................... 28
        4.2.2  Understanding Server Selection Dynamics ................ 29
        4.2.3  ISP-Friendliness ....................................... 31
        4.2.4  An ISP View of ISP-Friendliness Violation ............. 32
        4.2.5  Discussion ............................................. 33
   4.3  Energy-Aware Capacity Provisioning ............................. 33
   4.4  Implication and Motivation ..................................... 34
   References ......................................................... 34

**5  Capacity Provisioning for Video Content Delivery** .................. 35
   5.1  Problem Statement .............................................. 35
        5.1.1  The Network Model ...................................... 35
        5.1.2  Cost Function .......................................... 36
        5.1.3  Problem Formulation .................................... 36
   5.2  Capacity Provisioning Algorithm ................................ 37
        5.2.1  Characteristics of CDN Workload ........................ 38
        5.2.2  Estimating Global Capacity Lower Bound ................. 39
        5.2.3  Optimal ISP Capacity Allocation for Cost Saving ....... 40
        5.2.4  Allocating ISP Capacities with Reduced
               Server Switches ........................................ 41
   Appendix ........................................................... 42
   References ......................................................... 43

**6  Performance Evaluation** .......................................... 45
   6.1  Experiment Setup ............................................... 45
   6.2  Evaluation and Comparison ...................................... 47
        6.2.1  Overall Performance .................................... 47
        6.2.2  Performance from ISP Perspective ....................... 50
        6.2.3  Influence of Erroneous Workload Predictions ........... 52
        6.2.4  Influence of View Session Lengths ...................... 52
        6.2.5  Influence of Power and Bandwidth Prices ............... 54
   References ......................................................... 55

**7 Concluding Remarks** ....................................... 57
   7.1 Conclusions ........................................ 57
   7.2 Future Research Directions ............................ 57
   References .............................................. 59

# Chapter 1
# Introduction

In this chapter, we introduce the background of Content Delivery Networks (CDNs), and discuss the challenges and key problems in designing and running a CDN.

## 1.1 Background

With the advance of information and communication technologies, content services have become more and more popular on the Internet. Studies show that in 2013, 5 exabytes of contents were produced on the Internet every day [1]; in every minute of 2015, 2,460K pieces of contents were shared on Facebook, 277K messages were tweeted on Twitter, 216K new photos and 72 h of new videos were uploaded to Instagram and YouTube respectively [2]. Among the various types of Internet contents, videos are particularly important, as according to a white paper by Cisco [3], by the year 2020, video will dominate the Internet by accounting for as high as 82% of the global Internet traffic.

To cope with the accelerating speed of content production and consumption, Content Delivery Networks (CDNs), which enable a persistent and reliable content delivery to global users, have been proposed and rapidly developed on the Internet in the last decade. Currently, CDNs are serving a large fraction of the Internet contents, including web objects (text, graphics, scripts), downloadable objects (media files, software, documents), applications (e-commerce, portals), live streaming media, on-demand streaming media, and social networks.

Simply speaking, a CDN is an overlay network of servers that are deployed at many geographical locations. Instead of downloading contents from the original server, end users access the contents from the CDN servers that are approximate to them, and the CDN is obligated to provide the contents with high availability and performance [4]. Figure 1.1 shows the sever deployment of Limelight, a leading global CDN provider, as an example. One can see from the figure that to provide high available and high performance content services globally, a CDN needs to place its servers at tens or even hundreds of locations all around the world.

© The Author(s) 2017
Y. Tian et al., *Internet Video Data Streaming*,
SpringerBriefs in Computer Science, DOI 10.1007/978-981-10-6523-1_1

**Fig. 1.1** Infrastructure of Limelight CDN, data from [5]

Cyber giants like Google, Facebook, Microsoft, and Tencent have built their own dedicated CDNs for distributing contents to their enormous users. For example, Google's content distribution infrastructure is composed of a few core data centers and a large number of edge cache servers (referred to as Google Global Cache, GGC), and uses the infrastructure for serving various contents, including YouTube videos, Gmail mails, Google Play apps etc. [6]. On the other hand, commercial CDNs such as Akamai, Limelight, and ChinaCache provide content distribution services to third-party content providers. For example, recent studies reveal that Netflix, the leading subscription-based video streaming service provider in North America, outsources its video delivery to Akamai, Limelight, and Level3 [7, 8].

Note that although CDNs have been widely deployed and have carried most of the content traffics on the Internet, however, they are proprietary systems designed and operated by private enterprises, and there is no standard way for constructing and running a CDN. In the following, we will discuss the objectives of a CDN network, and analyze the key problems for fulfilling these missions.

## 1.2   CDN Objectives

### 1.2.1   Meeting Service Level Agreement

The primary objective for a CDN network is to provide content distribution services that satisfy the Service Level Agreement (SLA) between the CDN provider and the content provider. CDN SLA focuses on two aspects in general, content availability and delivery performance. Content availability is usually measured in term of "nines", which means the fraction of the content requests that are successfully served by CDN, defined as the following:

$$\text{content availability} = \frac{\text{content requests served by CDN}}{\text{total content requests}}$$

For example, availability of five nines means that the content is accessible in 99.999% of time, or the downtime is equal to 5.26 min per year or 25.9 s per month. Most CDNs provide availability between three nines and five nines, depending on the sensitivity of the content.

The content delivery performance metrics are measured in latency and throughput. Latency is critical for web-based content service, for example, Amazon reported that every 100 ms of latency in loading web pages leads to a 1% sales reduction; Bing found that a 2-s slowdown changed queries-per-user by $-1.8\%$ and revenue-per-user by $-4.3\%$ [9]. For video contents, users are expecting to experience at the highest bitrate that their last-mile connectivity, desktop, or device would allow, and with the applications such as 4K video and virtual reality becoming more and more popular, CDNs are expected to provide higher and higher throughputs. Note that to meet the throughput SLA, a CDN not only needs to provide sufficient bandwidth at its content servers, but more importantly, it must find the paths on the Internet that can sustain the high end-end throughput.

### 1.2.2 Reducing Traffic Expense

Another important mission for a CDN network is to reduce its operating expense (OPEX). There are two major components in a CDN's OPEX, the traffic expense and the energy expense. A CDN's traffic expense is largely determined by its relationship with the Internet Service Providers (ISPs): some CDNs, such as Limelight and Google, have built their private backbone networks that interconnect their data centers, and such a CDN backbone is peering with as many eyeball and transit ISPs as possible. If most of the CDN's traffic traverses the settlement-free peering links, considerable cost can be saved. For example, benefited from the peering between Google and many ISPs, it is estimated that YouTube, which is the largest video-sharing website run by Google, has a traffic cost close zero [10].

However, some other CDNs, which choose to deploy their server clusters at the ISP data centers, pay for their traffics to the ISPs in which the CDN servers reside. The payment is for compensating the traffic cost that the ISP pays to its provider ISPs. Currently, an ISP charges its customer ISP based on the traffic volume that the customer sends and receives within a charging period, e.g., one month. There are two popular ways to determine the charging volume: 95-percentile and total volume. In the former approach, traffic volume for each 5-min is recorded for a month and the 95-percentile over the sorted values is used as the charging volume. In the latter approach, the total traffic volume within a month is the charging volume. ISPs determine the final bill using a charging function with the computed charging volume. Note that the traffic cost paid by CDN will be eventually imposed on the content providers, and become a major source of the provider's business expense. For example, it is estimated that Netflix, which relies on three commercial CDNs for video delivery, paid about 50 million dollars to the CDNs in 2011 [11].

### *1.2.3  Reducing Energy Expense*

As an Internet-scale system comprising hundreds of thousands of servers, a CDN consumes a lot of energy, which has already become a significant source of its OPEX. For example, a large distributed platform with 100,000 servers will expend roughly 190,000 MWH per year, which is enough to sustain more than 10,000 households. In 2014, the total data centers' power consumption in USA was 2% of the total power consumption of the country, which has grew by 24% in 5 years. Furthermore, with the deployment of new services and the rapid growth of the Internet, the energy consumption of data centers is expected to grow at a rapid pace of more than 15% per year in the near future [12].

Note that in addition to reducing the monetary cost, reducing the energy consumption of a CDN is also very necessary for protecting the global environment, as it is estimated by the year 2020, the Green House Gas (GHG) emissions of the ICT sector are expected to reach 1.43 Gtons carbon dioxide equivalent ($CO_2e$) [13].

## 1.3  Challenges in CDN

It is not easy for a CDN to simultaneously accomplish the above mentioned objectives, and some objectives are contradicting to each other. For example, to improve content availability, a CDN may choose to serve with redundant servers, but over-provision leads to additional energy consumption. For another example, when forwarding content requests to servers, a CDN may face a dilemma between local servers with limited service capacity and remote data centers that have sufficient bandwidth and better Internet connectivity. In the following, we analyze the key problems in designing and operating a CDN.

### *1.3.1  System Architecture*

As a proprietary system upon the Internet, a CDN may follow different architecture design philosophies. One representative CDN design is to enter deep into the ISPs. In such a CDN, content servers are deploy at point-of-presences (PoPs) and third-party data centers in as many ISPs as possible. The rationale behind the design philosophy is the traditional pyramid ecosystem of the Internet connectivity, that is, the global

Internet is formed with tier-1 ISPs at the root of the Internet, and end users access the Internet from the numerous eyeball ISPs that directly or indirectly connect to tier-1 networks. The pyramid Internet connectivity ecosystem leads to highly fragmentation of the Internet content access—the top 45 ISPs combined account for only half of user access traffic, and the accesses drop off dramatically along the ISP index [4]. By placing content servers deep into the ISP networks, a CDN can deliver contents from the servers that are proximate to content requesting users, thus provide good delivery performances regarding the latency and throughput. One problem for the "deep-into-ISP" design is the management complexity, as the CDN is highly distributed with up to thousands of server clusters scattered around the world. Another issue is that the CDN may need to pay for its traffic cost to the ISPs in which its server clusters reside in, which increases the CDN's OPEX. A well-known example of the "deep-into-ISP" CDN architecture is Akamai [4], and nearly all the Chinese CDNs also follow this design philosophy [14].

The other design philosophy is to bring ISPs to home, that is, a CDN builds large data centers at key locations and connects these data centers with its high performance private backbone network. Usually a CDN of this type has its own AS(es), and connects transit and eyeball ISPs at key locations that are close to the ISP PoPs. The CDN design philosophy is motivated from the recent observations that majority of the inter-domain traffics on today's Internet flow directly between large content providers, data center/CDNs and consumer networks, and the global Internet becomes flatter with ISP networks peering with each other much more densely, due to the reasons such as the emerging Internet Exchange Points (IXPs) [10]. A representative commercial CDN of this type is Limelight [15], and Google also follows this design philosophy in its content service infrastructure [6].

### 1.3.2 Server Selection

One of the most important network mechanism for a CDN is server selection, that is, to select an appropriate content server or server cluster for accommodating end user's content request. A CDN makes its server selection decisions based on combination of many factors, such as latency [16], performance, [17], service capacity [18], etc., and different CDNs have different priorities in these factors.

Many CDN networks adopting the "deep-into-ISPs" architecture design employ an ISP-friendly server selection policy, that is, the CDN prefers to redirect a client's content request to servers in a same ISP of the client, or the ISPs that peer with the client's ISP, so as to avoid the cross-ISP traffic costs [14].

Existing CDN server selection mechanisms usually assume a constant service capacity. That is, the number of clients that a CDN's server cluster or data center can handle do not change over time. However, as clusters or data centers' capacities are dynamically "right sized", as we will discuss in the next subsection, a CDN's server selection problem should be jointly considered with the capacity provisioning problem, so as to optimize the CDN's overall expense [14].

### 1.3.3   Capacity Provisioning

The possibility of reducing a CDN's energy cost comes from two facts. First, study shows that a server's power consumption increases almost linearly with CPU utilization, while an idle server consumes up to 66% of the peak power, because even when a server is not loaded with user tasks, the power needed to run the OS and to maintain hardware peripherals, such as memory, disks, master board, PCI slots, and fans, is not negligible [19]. Second, the content requests arriving to a CDN exhibit a strong diurnal pattern, and vary significantly due to day, night, weekday, weekend, and holidays [20].

Based on the observations, recent studies [20, 21] propose to reduce a data center or a CDN's energy consumption by "right-sizing" its service capacity. The general idea is that the CDN chooses to hibernate a subset of its servers when the service requests can be satisfied with the remaining servers (e.g., during the mid-night hours), and awaken some sleeping servers when service requests start to increase (e.g., during the morning and evening hours).

Unfortunately, there exists a conflict between simultaneously saving a CDN's energy cost and avoiding its cross-ISP traffic. On the one hand, for saving the energy cost, we can keep a minimum number of the servers as long as the CDN's SLA is met. On the other hand, to avoid the cross-ISP content deliveries, we need to minimize the chance that the workload demand from an ISP exceeds the service capacity provisioned by the clusters in that ISP, and has to be accommodated by servers from other ISPs. Since workload fluctuates over time, the more service capacity we have provisioned, the less likely we will incur cross-ISP video deliveries. Apparently, for reducing the overall cost, the inherent conflict between the energy efficiency and the ISP-friendliness requires people to carefully plan the service capacities of the CDN clusters.

## 1.4   Overview of the Book

In this monograph, we consider both the energy and the cross-ISP traffic cost for a video streaming CDN. We present a capacity provisioning algorithm that is cost-aware by dynamically planning service capacities of the CDN clusters in numerous ISPs. Our work is motivated by the measurement study on the CDN infrastructure of Youku, which is the largest video site in China. Using workload derived from real-world traces and applying actual bandwidth and power price parameters, we show with experiments that our solution can balance a CDN's energy and bandwidth expenses, and significantly reduces its overall operating cost. In addition, our approach avoids unnecessary switches that toggling servers into and out of the power-saving mode, therefore can be practically applied on today's video streaming CDNs.

The monograph is organized as follows. Chapter 2 introduces the key issues, design choices, and performances of large-scale CDNs in real world. In Chap. 3, we discuss the representative techniques for saving energy consumed by server clusters, data centers, and CDNs. Chapter 4 presents the measurement study on the CDN infrastructure of Youku, and reveals the inherent conflict between the energy-aware capacity provisioning and the ISP-friendly server selection policy. In Chap. 5, we propose the cost-aware capacity provisioning algorithm that dynamically plans the CDN's service capacities in its clusters for saving the overall operating cost. Chapter 6 evaluates the proposed CDN capacity provisioning algorithm with workload derived from real-world measurement and actual price parameters. Finally, we discuss the future research directions in Chap. 7.

# References

1. Gunelius, S.: The data explosion in 2014 minute by minute—infographic. https://aci.info/2014/07/12/the-data-explosion-in-2014-minute-by-minute-infographic/ (2014)
2. James, J.: Data never sleeps 4.0. https://www.domo.com/blog/data-never-sleeps-4-0/ (2016)
3. Cisco visual networking index: forecast and methodology, 2015–2020. Cisco White Paper. http://www.cisco.com/c/en/us/solutions/collateral/service-provider/visual-networking-index-vni/complete-white-paper-c11-481360.html (2016)
4. Nygren, E., Sitaraman, R.K., Sun, J.: The Akamai network: a platform for high-performance internet applications. ACM SIGOPS Oper. Syst. Rev. **44**, 2–19 (2010)
5. Limelight global infrastructure. https://www.limelight.com/global-infrastructure/
6. Google Edge Network. https://peering.google.com/#/infrastructure
7. Adhikari, V.K., Guo, Y., Hao, F., Varvello, M., Hilt, V., Steiner, M., Zhang, Z.-L.: Unreeling netflix: understanding and improving multi-CDN movie delivery. In: Proceedings of IEEE INFOCOM, Orlando, FL, USA, pp. 1620–1628 (2012)
8. Adhikari, V.K., Guo, Y., Hao, F., Hilt, V., Zhang, Z.-L., Varvello, M., Steiner, M.: Measurement study of Netflix, Hulu, and a tale of three CDNs. IEEE/ACM Trans. Netw. **23**(6), 984–1997 (2015)
9. Hamilton, J.: The cost of latency. http://perspectives.mvdirona.com/2009/10/the-cost-of-latency/ (2009)
10. Labovitz, C., Iekel-Johnson, S., McPherson, D., Oberheide, J., Jahanian, F.: Internet inter-domain traffic. In: Proceedings of ACM SIGCOMM, New Delhi, India, pp. 75–86 (2010)
11. Rayburn, D.: Netflix's streaming cost per movie drops 50% from 2009, expected to spend $50M in 2011. http://blog.streamingmedia.com/2011/03/netflixs-streaming-costs-drop-50-from-2009-expected-to-spend-50m-in-2011.html (2011)
12. Sverdlik, Y.: Here's how much energy all US data centers consume. http://www.datacenterknowledge.com/archives/2016/06/27/heres-how-much-energy-all-us-data-centers-consume/ (2016)
13. GeSI: Smart 2020: enabling the low carbon economy in the information age. http://gesi.org/article/43 (2008)
14. He, H., Zhao, Y., Wu, J., Tian, Y.: Cost-aware capacity provisioning for internet video streaming CDNs. Comput. J. **58**(12), 3255–3270 (2015)
15. Huang, C., Wang, A., Li, J., Ross, K.-W.: Measuring and evaluating large-scale CDNs. Technical Report, Microsift Research, MSR-TR-2008-106 (2008)
16. Torres, R., Finamore, A., Kim, J.R., Mellia, M., Munafo, M.M., Rao, S.: Dissecting video server selection strategies in the YouTube CDN. In: Proceedings of ICDCS, Minneapolis, MN, USA, pp. 248–257 (2011)

17. Triukose, S., Wen, Z., Rabinovich, M.: Measuring a commercial content delivery network. In: Proceedings of WWW, Hyderabad, India, pp. 467–476 (2011)
18. Adhikari, V.K., Jain, S., Zhang, Z.-L.: YouTube Traffic dynamics and its interplay with a tier-1 ISP: An ISP perspective. In: Proceedings of IMC, Melbourne, Australia, pp. 431–443 (2010)
19. Meisner, D., Gold, B.T., Wenisch, T.F.: PowerNap: eliminating server idle power. In: Proceedings of International Conference on Architectural Support for Programming Languages and Operating Systems (ASPLOS), Washington, DC, USA, pp. 205–216 (2009)
20. Mathew, V., Sitaraman, R.K., Shenoy, P.: Energy-aware load balancing in content delivery networks. In: Proceedings of IEEE INFOCOM, Orlando, FL, USA, pp. 954–962 (2012)
21. Lin, M., Wierman, A., Andrew, L.L.H., Thereska, E.: Dynamic right-sizing for power-proportional data centers. IEEE/ACM Trans. Netw. **21**(5), 1378–1391 (2013)

# Chapter 2
# Content Delivery Networks and Its Interplay with ISPs

This chapter surveys measurement studies on Content Delivery Networks (CDNs) in real world. Several representative CDNs with different architecture designs, client mapping policies and redirection mechanisms are examined, and the interplay between CDNs and ISPs are further discussed.

## 2.1 CDNs in Real World

### 2.1.1 Akamai

**System Architecture**. Huang et al. [1] carry out a measurement study on the large-scale CDN of Akamai. The authors take two steps to chart a CDN network. In the first step, they find out all the Canonical NAMEs (CNAMEs) that are used by the target CDN's customers, by resolving 16 million web hostnames. A total number of 282,700 unique open local DNS (LDNS) servers are employed as vantage points in their measurement study. In the second step, the authors query the open LDNS servers all around the world with the CNAMEs, for discovering the Akamai content servers.

With the proposed methodology, the authors have successfully discovered 27,000 Akamai content servers, which the authors believe that complete the Akamai global network. The content servers are distributed in 65 countries in the world, and are in 656 Autonomous Systems (ASes). The authors also find that Akamai employs a two-tier DNS infrastructure with 6,000 name servers for resolving the second level domain names, and they conjecture that at least one name server is run in each distributed cluster of Akamai. The measurement result shows that Akamai adopts "deep-into-ISPs" architecture design in its CDN.

Researchers from Akamai later reveals its system architecture in [2], which is composed of an edge server platform, a mapping system, a communication and control system, and a data collection and analysis system. Akamai's edge servers are

© The Author(s) 2017
Y. Tian et al., *Internet Video Data Streaming*,
SpringerBriefs in Computer Science, DOI 10.1007/978-981-10-6523-1_2

responsible for processing end user requests and serving the requested contents, as well as for acting as intermediaries in the content delivery overlay network.

The mapping system of Akamai has two objectives, scoring and real-time mapping. In scoring, the system creates a topological map capturing the state of connectivity across the entire Internet by dividing the Internet into equivalence classes of IP addresses and represents how well they connect to each other. In real-time mapping, a top-level map first selects a preferred edge server cluster for each equivalence class of end users, according to the scoring result; then a low-level map within a cluster directs content request to a specific server.

Akamai views a content delivery network as a virtual network over the Internet. For video content delivery, Akamai organizes its edge servers as a tiered cache system, which can effectively improve the video delivery performance by reducing the number of content requests back to the origin server. For live video streaming, the transport system transports the stream's packets from the entry point to a subset of edge servers that require the stream using a publish-subscribe model.

**Mapping System**. Recently, Chen et al. [3] elaborate Akamai's mapping system, which is responsible for making real-time server selection decisions for user clients. They consider the conventional NS-based mapping system previously employed by Akamai as computing the following complex time-varying function:

$$Map_t : \Omega_{Internet} \times \Omega_{Akam} \times Domain \times LDNS \rightarrow IPs \qquad (2.1)$$

where $\Omega_{Internet}$ is the current state of the global Internet and $\Omega_{Akam}$ is the current state of the Akamai CDN. The two inputs are obtained through network measurement. *Domain* is the domain name of the content accessed by the client and *LDNS* is local recursive DNS server that makes the name resolution request on behalf of the client. A problem for the NS-based mapping is that server selection decisions are made based on the identity of the client's LDNS rather than that of the client itself, and some clients are far away from their LDNS servers, due to the wide usage of the public third-party name service providers such as Google DNS [4] and OpenDNS [5]. As a result, the NS-based mapping may select content servers that are distant from the clients and cause high latencies in downloading the contents.

Chen et al. [3] propose to exploit the recently proposed EDNS0 client-subnet extension, where LDNS forwards a prefix of the client's IP in name resolution requests, to assist the CDN to select servers. Conceptually, with the EDNS0 extension, the mapping system computes the following time varying function:

$$Map_t : \Omega_{Internet} \times \Omega_{Akam} \times Domain \times Client \rightarrow IPs \qquad (2.2)$$

where *Client* is the IP prefix of the client.

Akamai deploys the end-user mapping system around the world in the first half of 2014. By comparing the mapping results before and after the deployment, the authors find that end-user mapping provides significant performance benefits. In particular, for clients who use public resolvers, the end-user mapping achieves an eight-fold decrease in mapping distance, a two-fold decrease in RTT and content download time, and a 30% improvement in the time-to-first-byte.[1]

**Performance**. Triukose et al. [6] study performance of the Akamai CDN through an active measurement. The authors focus on the performance in terms of download throughput. By carefully constructing requesting URLs, the authors measure and compare the throughputs under the cases of cache hit, cache miss and directly downloading from the origin server. The comparison result shows that Akamai effectively improves the web performance, as in 67 and 41% of the cases, the CDN delivery is at least five times faster than cache-miss and origin delivery respectively.

The authors also examine Akamai's server selection decisions, and find that the CDN rarely selects the best server cluster with the highest download throughput, but it successfully avoids the worst ones. In roughly 75% of the cases, the Akamai-selected server outperformed half of the alternatives, suggesting that there is still space for improvement.

Motivated by the observations, the authors consider consolidating the existing infrastructure of the Akamai CDN into fewer clusters. With simulation experiments driven by the measurement data, it is found that clients would not see noticeable performance difference if the servers were further consolidated into 40 clusters, given that the CDN can always direct clients to the best servers.

### 2.1.2 Limelight

Huang et al. [1] also employ the same platform for measuring Akamai to chart the Limelight CDN. The authors find that Limelight uses IP anycast to announce its name servers, and reuses its content servers as DNS servers as well. The 4,100 discovered content servers are located in only 19 locations, making Limelight a representative CDN following the "bringing ISPs at home" design philosophy. In addition, it is observed that Limelight's anycast routing does not redirect clients to the closest data centers more than 37% of the time, and the redirection is beyond the top three data centers in about 20% of the time.

The authors then compare Limelight with Akamai regarding the CDN health and the delay performances. Limelight is observed to have better server availability and longer uptime, suggesting that comparing with Akamai, Limelight servers are easier to maintain. For evaluating the delay performance, the authors propose a revised King approach to measure the delay from each LDNS vantage point to the CDN-selected server. The measure result shows that the delay performance has a positive correlation with the coverage of the CDN in a specific target geographic region, and Limelight has considerable longer delays than Akamai in Europe, Asia, Oceania, South America, and Africa, due to its fewer data centers deployed in these regions.

---

[1]Time-to-first-byte, or TTFB, refers to the duration from when the client makes a HTTP request for the base web page to when the first byte of the requested web page was received by the client.

### 2.1.3  Amazon CloudFront

Bermudez et al. [7] study Amazon's Web Service (AWS) provided by EC2, S3 and CloudFront. The authors employ one-week traffic collected by Tstat, a traffic monitoring tool, at two ISP PoPs and their campus network. A total number of $6M$ TCP connections between clients and AWS servers were examined in their passive measurement study.

The authors find that for EC2 and S3, workload distributed among different data centers are unbalanced, with 85% European clients being redirected to a data center in US, leading to long latencies, high network cost, and low throughput. The authors find that the CloudFront CDN deploys its servers in 33 locations, and by directing clients to close servers, the CDN shows excellent web performance comparing with EC2 and S3.

### 2.1.4  Google

**Infrastructure**. Calder et al. [8] carry out a study of Google's web content delivery infrastructure over a ten-month period. The authors have proposed a suite of methods for mapping Google's infrastructure. More specifically, the EDNS-client-subnet extension is exploited to enumerate addresses of Google's front-end servers, and has collected 28,793 front-end IP addresses. The authors propose a novel technique named client-centric geolocation (CCG) to approximate the location of a server by the geographical mean of the clients that it serves. Finally, the authors measure the RTTs of the front-ends to PlanetLab node landmarks, and cluster them into CDN nodes with the OPTICS algorithm [9].

Armed with the methodologies, the authors present a detailed analysis on Google's web serving infrastructure and its expansion in ten month. It is observed that instead of serving clients from front-end servers on its own backbone network, Google expands its infrastructure to the ASes in lower tiers of the AS hierarchy. As a result, although Google still directs majority of clients to servers in its own network, an additional 8% of clients are served off its network; moreover, the expansion makes clients to be directed to servers that are substantially closer to them. The observation suggests that Google has evolved from a typical "bringing ISP to home" CDN to a hybrid one, as in addition to maintaining its own backbone network and data centers, Google also deploys its server clusters deep into the small eyeball ISPs.

**Front-end Server Placement**. Chen et al. [10] consider the front-end server placement in dynamic content distribution. The authors consider Google and Bing as two cases, and perform extensive network measurement on their CDNs. The authors propose a simple model-based inference framework to indirectly measure and quantify the "front-end to back-end fetching time" in the two CDNs, and point out that there is a distance threshold within which placing font-end servers further closer to users is no longer helpful; instead, the end-to-end performance is now determined by the front-end to back-end fetching time, which is determined by the query processing time at back-end data centers and the delivery time between the back-end data centers and the front-end servers. The authors also show that Google outperform Bing in content fetching time, despite that it's front-end servers are slightly farther from the clients.

**Path Optimization**. Krishnan et al. [11] diagnose the latency inflation of the end-to-end paths between clients and Google's CDN nodes. Google's server selection decisions are made based on latency, more specifically, Google gathers a map of latencies from CDN nodes to IP address prefixes. The latency is collected by redirecting a client to a random node, measuring the RTT, and using it as representative of the client's prefix. With RTT measurements gathered over time, and refreshed periodically, Google determines the closest CDN node for each IP address prefix, and redirects clients in the prefix to the closest node.

For evaluating effectiveness of the latency-based server selection, the authors examine the RTT data measured by 13 nodes in Google's CDN to the 170 K network prefixes they serve in one day. The measurement result shows that although each client is served by the CDN node measured to have the lowest latency to the client's prefix, in 40% cases, the measured RTTs are greater than 400 ms. In other words, the latencies experienced by clients are poor. By further analyze the RTT measurement data and probing prefixes with traceroute, the authors find that there are two major causes for the latency inflation: First, some clients have circuitous routes to or back from the CDN node serving them; Second, TCP connections to most clients are impacted by significant queueing delays.

The authors design and implement *WhyHigh*, a network diagnosis system, to identify the client prefixes that suffer latency inflation and pinpoint the causes for the instances of latency inflation. Inflated prefixes at a node are identified by comparing the minimum RTT measured at the node across all connections to the prefix with the minimum RTT measured at the same node across all connections to clients within the prefix's region. A prefix is declared to be inflated if the RTT difference is greater than 50 ms. For each inflated prefix instance, WhyHigh attempts to pinpoint the set of causes. The plausible causes include lack of peering between the prefix's ISP and Google, limited bandwidth of the peering link, ISP routing misconfiguration, and traffic engineering by ISP.

WhyHigh has identified and diagnosed a total number of 11,862 inflated prefixes globally, and among them, 3,776 instances are due to the lack of peering. The information provided by WhyHigh has helped Google to work with ISPs in the region and significantly improve path latencies.

### 2.1.5 Bing

Calder et al. [12] study the CDN of Bing, which relies on anycast to direct clients to front-end servers. Anycast is a routing strategy where the same IP address is announced from many locations throughout the Internet. Then BGP routes clients to one front-end location based on the best BGP path. The authors combine Bing server logs and Javascript-based active measurements to discover front-end servers of the Bing CDN, and measure the latencies between clients and different front-end servers, including the anycasted servers and the top-ten closest servers. The measurement results show that anycast usually performs well despite the lack of precise centralized control, but it directs roughly 20% of clients to a suboptimal front-end server.

The authors also find that the anycast server selections are quite stable, as clients continue going to the same front-end servers over time. The authors propose a simple prediction scheme, and shows with simulation that traditional and recent DNS techniques can improve performance for many of the clients who experience suboptimal anycast routing, thus the DNS-based redirection can be applied for only a small subset of clients that suffer poor server selections, while leaving the others to anycast.

### 2.1.6  YouTube

**Load Balancing**. Adhikari et al. [13] conduct an extensive and in-depth study of traffic exchanged between YouTube data centers and its users, as seen from the perspective of a tier-1 ISP in Spring 2008 after YouTube was acquired by Google but before Google did any major restructuring of YouTube. The authors employ the flow-level data related to YouTube that is collected from 44 ISP PoPs in their study, and find that YouTube employs a proportional load balancing strategy to split traffic among the seven data centers, which is very different from latency or location-based strategies that are usually observed, and the proportionality seems to be determined by the "size" of the data centers. The authors also find asymmetric "early-exit" routing paths for client-to-YouTube and YouTube-to-client traffics.

The authors perform a "what if" analysis on the scenario that YouTube adopts a location-based server selection strategy by directing clients to geographical nearby data centers, and show that under such a strategy, the overall video download latency can be reduced, but the workload among the datacenters will become unbalanced.

**Server Selection**. Torres et al. [14] conduct a detailed study of the YouTube CDN, and seeks to understand the mechanisms and policies for determining which data centers users download video from. The authors employ week-long datasets simultaneously collected from five networks, including ISPs and campus networks in 2010, and apply the delay-based CBG method [15] to geolocate the YouTube video servers. A total number of 33 data centers are identified. By analyzing the traffic datasets, the authors find that the data center that provides most of the traffic is also the data center with the smallest RTT for each dataset.

The authors also investigate the video downloading from non-preferred data centers, and find that a variety of causes, including load balancing across data centers, variations across DNS servers within a network, alleviation of hotspots due to popular video content, and accesses of sparse video content that may not be replicated across all data centers, are responsible for the video downloading from non-preferred data centers.

### 2.1.7 Netflix and Hulu

Adhikari et al. [16, 17] investigate the systems of Netflix and Hulu, the leading Over-the-Top (OTT) video service providers in the US and Canada. The authors find that both providers employ the same set of three CDNs, namely Akamai, Limelight, and Level3, to distribute video contents. For CDN selection, Netflix assigns a fixed CDN preference order for each user account, and a Netflix client stays with a same CDN as long as possible even if it has to degrade the playback quality level. For each client, Hulu chooses the preferred CDN for each video based on a latent distribution, and when the selected CDN has degraded bandwidth, Hulu adjusts the playback bit rates and continues to use the same CDN server as long as possible.

Motivated by the fact that the CDN preference and selection strategies employed by Netflix and Hulu are agnostic of the varying network conditions of the CDNs, the authors propose a measurement-based adaptive CDN selection strategy and a multiple-CDN-based video delivery strategy. The strategies meet the business constraints between the CDNs and the provider, and can significantly increase users' average available bandwidth in video streaming, by dynamically switching the streaming CDN according to real-time network conditions of the CDNs.

### 2.1.8 CDNs in China

Xue et al. [18] focus on the Chinese CDNs, which are different from global CDNs because of two reasons. First, the Chinese Internet is highly hierarchical with relatively few ASes [19]. Links between ASes are statically configured without using BGP, such a feature makes approaches like anycast fairly ineffective in China. Second, a global CDN usually cannot place content servers in China due to legal reasons, therefore must partner with Chinese CDNs.

The authors carry out a measurement-based study on a number of Chinese CDNs and global CDNs, and have the following observations: (1) all the Chinese CDNs employ the DNS rediection scheme, and tend to select a static set of servers; (2) Chinese CDNs provide customized content delivery service for web sites, leading to different latency performance when accessing these web sites, despite that they are served by a same CDN provider; (3) global CDNs who partner with Chinese CDNs can achieve significantly lower latencies than those who do not, because the contents are provided by servers inside China. The authors also discuss the possible reasons of poor server selections, which should be useful for optimizing the CDNs in China.

## 2.2  Interplay Between CDN and ISP

### 2.2.1  ISP Prospective

Poese et al. [20] consider the inherited conflict between ISP and CDN. More specifically, the authors consider the problem that today's CDN optimizes its traffic flows for minimizing its operational cost under the SLA constraint, the traffic scheduling result may be sub-optimal to the ISPs, and impose a heavy burden on them. For example, if the ISP changes its routing, e.g., for the purpose of traffic engineering, the content delivery network may re-optimize this delivery strategy and change the traffic matrix, which may invalidate the traffic engineering choice of the ISP [21].

To overcome this impairment, the authors propose a solution where the ISP offers a *Provider-aided Distance Information System* (*PaDIS*). PaDIS works as a location recommendation service operated by ISP. When a user, a CDN, a DNS resolver, or any other entity queries PaDIS by submitting a list of possible IP addresses and a source, PaDIS ranks the submitted IP addresses according to its metrics such as distance within the Internet topology, path capacity, path congestion, path delay, etc. The authors evaluate PaDIS in real-world measurement experiments, and find that with PaDIS, end-user download time is significantly reduced, while the ISP regains the ability to perform traffic engineering by biasing CDN's server selection decisions.

### 2.2.2  CDN Prospective

Khare et al. [22] considers the CDN-ISP interplay from the CDN prospective. Concretely, CDNs pay the ISPs for using the underlying networks based on the traffic volume that CDN servers send and receive. If an ISP's operational cost goes up due to the increase of inter-domain payments, it will eventually be reflected in higher traffic charges to the CDNs. Therefore CDN and ISP have the common interest in reducing the ISP's inter-domain payments, which will reduce the operational costs for both of them. The authors propose that to economically incentivize a CDN to consider the underlying ISPs' routing preferences, an ISP should charge differently for content traffic depending upon the type of inter-domain links it traverses, and make this pricing information available to the CDNs.

The authors devise a novel server assignment mechanism, which seeks to minimize a CDN's payment to the ISPs by directing clients to servers in the "right" ISPs under the new traffic pricing model. By comparing with the nearest-available sever selection policy, the authors show that the proposed mechanism provides good content delivery performance for end users, and has significant savings on the CDN's traffic payment.

# References

1. Huang, C., Wang, A., Li, J., Ross, K.W.: Measuring and evaluating large-scale CDNs. Technical Report, Microsift Research, MSR-TR-2008-106 (2008)
2. Nygren, E., Sitaraman, R.K., Sun, J.: The Akamai network: a platform for high-performance Internet applications. ACM SIGOPS Oper. Syst. Rev. **44**, 2–19 (2010)
3. Chen, F., Sitaraman, R.K., Torres, M.: End-user mapping: next generation request routing for content delivery. In: Proceedings of ACM SIGCOMM, London, UK, pp. 167–181 (2015)
4. Google Public DNS: http://developers.google.com/speed/public-dns/
5. OpenDNS: http://www.opendns.com/
6. Triukose, S., Wen, Z., Rabinovich, M.: Measuring a commercial content delivery network. In: Proceedings of WWW, Hyderabad, India, pp. 467–476 (2011)
7. Bermudez, I., Traverso, S., Mellia M., Munafo, M.: Exploring the cloud from passive measurements: the Amazon AWS case. In: Proceedings of IEEE INFOCOM, Turin, Italy, pp. 230–234 (2013)
8. Calder, M., Fan, X., Hu, Z., Katz-Bassett, E., Heidemann, J., Govindan R.: Mapping the expansion of Google's serving infrastructure. In: Proceedings of IMC, Barcelona, Spain, pp. 313–326 (2013)
9. Ankerst, M., Breunig, M., Kriegel, H., Sander, J.: OPTICS: ordering points to identify the clustering structure. ACM SIGMOD Record. **28**(2), 49–60 (1999)
10. Chen, Y., Jain, S., Adhikari V.K., Zhang Z.-L.: Characterizing roles of front-end servers in end-to-end performance of dynamic content distribution. In: Proceedings of IMC, Berlin, Germany, pp. 559–568 (2011)
11. Krishnan, R., Madhyastha, H.V., Srinivasan, S., Jain, S., Krishnamurthy, A., Anderson, T., Gao, J.: Moving beyond end-to-end path information to optimize CDN performance. In: Proceedings of IMC, Chicago, IL, USA, pp. 190–201 (2009)
12. Calder, M., Flavel, A., Katz-Bassett, E., Mahajan, R., Padhye, J.: Analyzing the performance of an anycast CDN. In: Proceedings of IMC, Tokyo, Japan, pp. 531–537 (2015)
13. Adhikari, V.K., Jain, S., Zhang, Z.-L.: YouTube traffic dynamics and its interplay with a tier-1 ISP: an ISP perspective. In: Proceedings of IMC, Melbourne, Australia, pp. 431–443 (2010)
14. Torres, R., Finamore, A., Kim, J.R., Mellia, M., Munafo, M.M., Rao, S.: Dissecting video server selection strategies in the YouTube CDN. In: Proceedings of ICDCS, Minneapolis, MN, USA, pp. 248–257 (2011)
15. Gueye, B., Ziviani, A., Crovella, M., Fdida, S.: Constraint-based geolocation of internet hosts. IEEE/ACM Trans. Network. **14**(6), 1219–1232 (2006)
16. Adhikari, V.K., Guo, Y., Hao, F., Varvello, M., Hilt, V., Steiner, M., Zhang, Z.-L.: Unreeling Netflix: understanding and improving multi-CDN movie delivery. In: Proceedings of IEEE INFOCOM, Orlando, FL, USA, pp. 1620–1628 (2012)
17. Adhikari, V.K., Guo, Y., Hao, F., Hilt, V., Zhang, Z.-L., Varvello, M., Steiner, M.: Measurement study of Netflix, Hulu, and a tale of three CDNs. IEEE/ACM Trans. Network. **23**(6), 984–1997 (2015)
18. Xue, J., Choffnes, D., Wang, J.: CDNs meet CN: an empirical study of CDN deployments in China. IEEE Access **5**, 5292–5305 (2017)
19. Tian, Y., Dey, R., Liu, Y., Ross, K.W.: China's Internet: topology mapping and geolocating. In: Proceedings of IEEE INFOCOM, Orlando, FL, USA, pp. 2531–2535 (2012)
20. Poese, I., Frank, B., Ager, B., Smaragdakis, G., Feldmann, A.: Improving content delivery using provider-aided distance information. In: Proceedings of IMC, Melbourne, Australia, pp. 22–34 (2010)
21. Jiang, W., Zhang-Shen, R., Rexford, J., Chiang, M.: Cooperative content distribution and traffic engineering in an ISP network. In: Proceedings of ACM SIGMETRICS, Seattle, WA, USA, pp. 239–250 (2009)
22. Khare V., Zhang, B.: Making CDN and ISP routings symbiotic. In: Proceedings of ICDCS, Minneapolis, MN, USA, pp. 869–878 (2011)

# Chapter 3
# Energy Management

This chapter reviews various techniques of improving energy efficiency in data centers and content delivery networks (CDNs), including dynamic voltage frequency scaling (DVFS), resource virtualization and migration, capacity right-sizing, and cost saving method using batteries.

## 3.1 Energy Saving for Data Center

### 3.1.1 DVFS-Based Energy Saving

Many CPUs have a DVFS (dynamic voltage frequency scaling) architecture that can dynamically scale processor voltage and frequency. Hotta et al. [1] exploit the DVFS feature and propose a profile-based power-performance optimization technique for energy saving. The proposed methodology is based on the observation that in parallel applications, a large amount of execution time may be spent for exchanging data between nodes in a cluster. When a node waits for communication from other nodes, its power consumption can be reduced by setting a lower frequency of the processor.

The authors propose to divide a program into regions. For each region $r_i$, if it is executed at frequency $f_i$, the authors use the energy delay product (EDP) $Er(r_i, f_i)$ to express its energy efficiency, and the energy saving problem becomes to minimize the overall energy delay product $Er(P)$ of the program $P$, as

$$Er(P) = \sum_{i=0}^{n} Er(r_i, f_i) + E_{trans}(r_i, f_i) \qquad (3.1)$$

where $E_{trans}(r_i, f_i)$ represents the summation of the transition overhead in terms of EDP to change to frequency $f_i$. A greedy algorithm is designed to find the frequency $F_i$ for each region $r_i$. The authors develop a power profile system named *PowerWatch* for measuring power consumption on each HPC node, and with the power profile

© The Author(s) 2017
Y. Tian et al., *Internet Video Data Streaming*,
SpringerBriefs in Computer Science, DOI 10.1007/978-981-10-6523-1_3

system, the authors experiment their proposed technique on two kinds of clusters with Turion and Crusoe processors. The experiment results show that the DVFS-based optimization is promising for improving energy efficiency of HPC clusters.

### 3.1.2  Energy Saving with Virtualization

In a cloud system, dynamic virtual machine (VM) relocation and consolidation can be used to migrate VMs away from underutilized servers, and transition idle servers into a lower power-state for energy saving. Feller et al. [2] present a holistic solution for VM resource utilization monitoring and estimation, and within their framework, the authors propose several energy management algorithms and mechanisms. More specifically, four types of scheduling policies are presented: placement, overload relocation, underload relocation, and consolidation. Placement policies are triggered event-based to place incoming VMs on nodes. Relocation policies are called when overload events arrive from nodes and seek to move VMs away from heavily loaded nodes. On the contrary, in case of underload, VMs are moved to moderately loaded nodes in order to create underutilized nodes and transition them into a lower power state. Finally, consolidation policies are called periodically to further optimize the VM placement of moderately loaded nodes. For overload and underload relocation, greedy algorithms are proposed; while for VM consolidation, which is an NP-hard problem, the authors employ a modified Sercon algorithm [3].

The authors implement their proposed algorithms in the *Snooze* VM management framework [4], and evaluate the energy and performance implications on 34 power-metered servers of the Grid'5000 experimentation testbed under dynamic web workloads. The results show that substantial energy savings can be achieved with only limited impact on web performance.

Liu et al. [5] develop a performance and energy model of VM live migration in a cloud environment. With theoretical analysis, the authors show that the energy consumption $E_{mig}$ of a VM migration increases linearly with the network traffic volume of VM migration, i.e.,

$$E_{mig} = a \times V_{mig} + b \qquad (3.2)$$

where $V_{mig}$ is the VM migration traffic. An algorithm, which takes system parameters including VM memory size, memory transmission rate, memory dirtying rate, and the threshold of dirty memory transmission as inputs, is developed for estimating $V_{img}$.

By applying linear regression technique on empirical data, the authors derive that $a = 0.512$ and $b = 20.165$. The authors also develop a prototype on Xen platform, in which the energy model is applied to select the VM that has the minimum cost to migrate each time. Experimental results show that the model-guided migration decisions can significantly reduce the migration cost by more than 72.9% at an energy saving of 73.6%.

### 3.1.3 Energy Saving with Capacity Right-Sizing

A data center's energy consumption can be further reduced based on two observations. First, a server's power consumption increases almost linearly with CPU utilization, while an idle server consumes up to 66% of the peak power [6]; Second, the workload arriving to a data center exhibit a strong diurnal pattern, and vary significantly due to day, night, weekday, weekend, and holidays, and such workload can be accurately forecasted.

Chen et al. [7] consider saving energy consumption of connection servers, which host a large number of long-lived TCP connections, for a cloud service. The authors seek to provision the minimal number of connection servers with the following concerns. First, each content server is subject to a maximum login rate and a maximum number of sockets it can host, due to the OS memory constraints and the fault tolerance consideration. Second, three types of errors, namely the Service Not Available (SNA) error, Server-Initiated Disconnection (SID), and Transaction Latency, should be avoid.

Two interdependent techniques: dynamic provisioning and load dispatching are proposed for saving energy consumptions of content servers. The dynamic provisioning technique dynamically turns on a minimum number of servers required to satisfy application-specific quality of service, based on the forecasted future workload; the load dispatching technique distributes current workload among the running servers, so that new login requests are routed to busy servers as long as the servers can handle them, and a small number of tail servers that have few connections are starved and being ready to shut down. With the experiments using Windows Live Messenger workload, the authors show that up to 30% of the energy can be saved without sacrificing user experiences.

Lin et al. [8] propose to "right size" cloud data center by turning off servers during predictable low workload period. An analytical model is developed to capture the major issues in data center right-sizing problem, including the cost associated with the increased delay from using fewer servers, the energy cost of maintaining an active server with a particular workload, and the cost incurred from toggling a server into and out of a power-saving mode. An optimal offline solution is analyzed, and the authors show that it exhibits a simple, "lazy" structure when viewed in reverse time. Based on their analysis, the authors propose a novel and practical online solution, and theoretically prove that it is 3-competitive, that is, the cost incurred by the algorithm is at most 3 times of the optimal offline solution. By evaluating with real-world data center workloads, the authors show that the proposed algorithm can potentially save 50% of the energy under workloads with a peak-to-mean ratio (PMR) as high as 5.

## 3.2    Energy Saving for CDN

### 3.2.1    Energy-Aware Load Balancing

Mathew et al. [9] also address the energy saving problem for CDN. More specifically, an energy-aware CDN should turn off content servers during periods of low
workload while seeking to balance three key design goals: maximize energy reduction, minimize the impact on client-perceived service availability (SLAs), and limit
the frequency of on-off server transitions to reduce wear-and-tear and its impact on
hardware reliability.

The authors propose energy-aware CDN load-balancing algorithms in two stages.
The local load balancing algorithm works within a CDN server cluster, and makes
decisions on hibernating and waking up servers based on past and current workload,
then the global energy-aware load balancing algorithm redistributes traffic across
clusters according to real-time cluster capacities. Evaluation using Akamai CDN
workload shows that by holding an extra 10% of the servers as live spares, the local
load balancing algorithm achieves a system-wide energy reduction of 55% and a
service availability of at least five nine's (99.999%), while incurring an average of
at most 1 transition per server per day. Although having limited impact on energy
reduction, the global load balancing algorithm can reduce 10–25% server transitions
by redistributing workload across proximal clusters.

### 3.2.2    Battery-Based Power Saving

A CDN network which deploys its server clusters at hundreds of locations around
the world usually rents servers or racks from third-party data centers, and pays each
data center for the amount of power supplied to their servers, not for the energy their
servers actually use. Under such a tariff model, Palasamudram et al. [10] propose to
use batteries to reduce both the required power supply and the incurred power cost
of a CDN.

The authors define two optimization problems, namely the power supply minimization (TPM) problem and the power cost minimization (PCM) problem, and
theoretically characterize the benefits that batteries provide and how these benefits
vary with power demand, battery characteristics, power prices, battery prices, battery
lifetimes, and server power proportionality. The author present linear programming
based solutions for the TPM and PCM problems, and use Akamai workload traces
for evaluating their effectiveness. The results show that up to 14% power savings
can be achieved, and the benefit would increase even more to 35.3% for perfectly
power-proportional servers. The cost savings, including the additional battery costs,
range from 13.26 to 33.8% as servers become more power-proportional.

# References

1. Hotta, Y., Sato, M., et al.: Profile-based optimization of power performance by using dynamic voltage scaling on a PC cluster. In: Proceedings of IEEE IPDPS, Rhodes Island, Greece, pp. 1–8 (2006)
2. Feller, E., Rohr, C., Margery, D., Morin, C.: Energy management in IaaS clouds: a holistic approach. In: Proceedings of IEEE International Conference on Cloud Computing, Honolulu, HI, USA, pp. 204–212 (2012)
3. Murtazaev, A., Oh, S.: Sercon: server consolidation algorithm using live migration of virtual machines for green computing. IETE Techn. Rev. 28(3), 212–231 (2011)
4. Feller, E., Rilling, L., Morin, C.: Snooze: a scalable and autonomic virtual machine management framework for private clouds. In: Proceedings of IEEE/ACM International Symposium on Cluster, Cloud, and Grid Computing, Ottawa, Canada, pp. 482–489 (2012)
5. Liu, H., Jin, H., Xu, C., Liao, X.: Performance and energy modeling for live migration of virtual machines. Cluster Comput. 16(2), 249–264 (2013)
6. Meisner, D., Gold, B.T., Wenisch, T.F.: PowerNap: eliminating server idle power. In: Proceedings of International Conference on Architectural Support for Programming Languages and Operating Systems (ASPLOS), Washington, DC, USA, pp. 205–216 (2009)
7. Chen, G., He, W., Liu, J., Nath, S., Rigas, L., Xiao, L., Zhao, F.: Energy-aware server provisioning and load dispatching for connection-intensive internet services. In: Proceedings of USENIX Symposium on Networked Systems Design and Implementation (NSDI), Renton, WA, USA, pp. 337–350 (2008)
8. Lin, M., Wierman, A., Andrew, L.L.H., Thereska, E.: Dynamic right-sizing for power-proportional data centers. IEEE/ACM Trans. Netw. 21(5), 1378–1391 (2013)
9. Mathew, V., Sitaraman, R.K., Shenoy, P.: Energy-aware load balancing in content delivery networks. In: Proceedings of IEEE INFOCOM, Orlando, FL, USA, pp. 954–962 (2012)
10. Palasamudram, D.S., Sitaraman, R.K., Urgaonkar, B., Urgaonkar, R.: Using batteries to reduce the power costs of internet-scale distributed networks. In: Proceedings of ACM Symposium on Cloud Computing (SoCC), San Jose, CA, USA, pp. 1–14 (2012)

# Chapter 4
# Cost Measurement for Internet Video Streaming

In this chapter, we carry out a measurement study on the CDN infrastructure of Youku, which is the largest video broadcasting and provider site in China and the second largest site in the world behind YouTube. In particular, we present an insightful analysis on its server selection behaviors. We find that Youku follows an ISP-friendly policy in general, but the CDN also violates the policy when balancing the excessive workloads from the ISPs. It is shown that for a Youku-like CDN, there is an inherent conflict between the energy-aware capacity provisioning, which right-sizes the CDN's service capacity for saving its energy expense, and the ISP-friendly server selection policy that seeks to avoid the cross-ISP video traffic.

We describe our measurement methodology and unveil the CDN design in Sect. 4.1; the CDN's server selection policy is analyzed in Sect. 4.2; in Sect. 4.3, we characterize the energy-aware capacity provisioning techniques that aim to improve the system's energy efficiency; we identify the inherent conflict between a CDN's energy efficiency and its ISP-friendly policy in Sect. 4.4, and discuss its implications as well.

## 4.1 Measurement Methodology and CDN Architecture

There are two objectives in our measurement study on Youku's CDN: (1) to unveil the CDN's system design; and (2) to collect a rich set of the server selection samples from various geographical locations and ISPs for enabling a policy analysis.

To achieve the two objectives, we exploit Youku's built-in server selection mechanism. Figure 4.1 demonstrates the major steps in this mechanism: When a client's web browser parses a web page in which the video is embedded, a static video URL like "http://f.youku.com/..." is retrieved (step 1–2). The client then queries its local DNS server for the name f.youku.com, and gets a CNAME reply

© The Author(s) 2017
Y. Tian et al., *Internet Video Data Streaming*,
SpringerBriefs in Computer Science, DOI 10.1007/978-981-10-6523-1_4

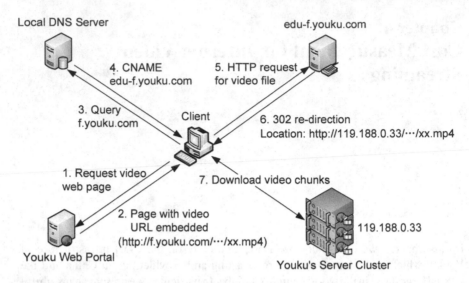

**Fig. 4.1** Demonstration of Youku's server selection steps

like edu-f.youku.com (step 3–4).[1] After the DNS resolution, the client sends out an HTTP GET request to the host binding the CNAME (step 5). However, the host doesn't have the requested video, but replies with an HTTP 302 re-direction message, which contains the IP address of the content server that actually hosts the video file (step 6). Finally, the user client follows the re-direction and downloads video chunks from the content server (step 7).

With the understanding of Youku's server selection mechanism, we can see that if we emulate a client's video request through an HTTP proxy, as the proxy queries its local DNS server and uses its own IP address when forwarding the request, Youku will select a server that is "optimal" for the proxy, and returns it's IP address in the 302 re-direction message via the proxy to our measurement agent. In other words, we can collect a *sample* on Youku's server selection decisions from the proxy. Furthermore, by probing through many HTTP proxies distributed on a wide range of geographical locations and ISP networks, a large number of the samples can be harvested for analyzing the CDN's server selection policy.

Based on the methodologies above described, we carry out a measurement study on Youku's CDN. To filter out the influence of the content availability on CDN's server selection decisions, in each probe we only request the "headline" video, that is, the video posted at the headline position on Youku's web portal. The video could be a breaking news, an important social event, or any other content that Youku wishes to promote. As it is under promotion, the CDN typically caches the video to the maximum extent.

---

[1] Youku's authoritative DNS server resolves the DSN query with different CNAMEs, and besides edu-f.youku.com, we have observed four other CNAMEs when probing from different ISPs and locations.

**Fig. 4.2**  Geographic distribution of Youku's server clusters

We carried out an 83-day measurement on Youku. In each day of the measurement, we collect more than one hundred HTTP proxies from various resources (e.g., websites, forums and blogs),[2] and use them to probe Youku's CDN. A total number of 2,997 distinct HTTP proxies were employed in our measurement. By using the Cymru IP to AS mapping tool [1] and applying the geolocating technique in [2], we find that the proxies are widely distributed in 267 cities and 55 ASes, covering nearly all the cities and ISPs in China. By probing from these proxies, we have found 759 distinct CDN server IP addresses in 43 cities and 17 ASes. We further group the server addresses that are in a same city and a same AS into a cluster, which we refer to as a CDN's *server cluster*. We have found a total number of 54 clusters and show their geographical distribution in Fig. 4.2.

From Fig. 4.2, one can see that Youku adopts a CDN design that deploys its server clusters at dozens of cities in China. In fact, unlike YouTube, which employs a few massive data centers [3], Youku does not possess any data center, but places its servers at the ISP or third-party data centers at as many locations and ISP networks as possible. Note that such a "deep-into-ISPs" design is representative, as it is found that many large scale CDNs in China and in the world, such as ChinaCache and Akamai, construct their networks in a similar way [4].

---

[2]These proxies were provided by individuals and organizations for different purposes, such as bypassing local ISP's control policies, or enabling anonymous web surfing, etc., examples can be found at http://www.cnproxy.com/.

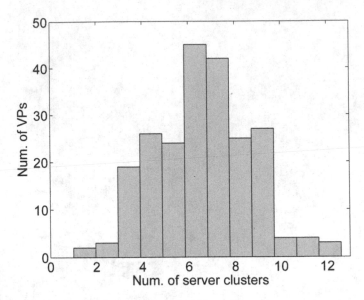

**Fig. 4.3** Histogram of server clusters selected for VPs

## 4.2   Server Selection Policy Analysis

We investigate Youku's server selection policy based on the samples collected from the proxies. Similar to the aggregation of the server clusters, we group the proxies that are in a same city and a same AS into a cluster, which we refer to as a measurement vantage point (or a *VP* for short). By filtering out the VPs with less than 15 samples, we have grouped 229 VPs, which are distributed in 142 cities and 32 ASes in China.[3] A total number of 24,533 server selection samples were collected from these VPs.

### 4.2.1   Server Selection Characteristics

We first examine the server clusters that were selected by Youku for the VPs. Figure 4.3 presents the histogram of numbers of the clusters selected for the VPs in the 83-day measurement. From the figure one can see that there are considerable *dynamics* in Youku's server selections, as for most VPs, more than one clusters were selected over time.

We then focus on the selection frequencies. For a VP $v$, we compute the selection frequency $f_v(k)$ of its $k$th most selected cluster $c_k^v$ as the ratio of the times that the cluster got selected. For all the VPs under study, we compute a *normalized cluster*

---

[3]The reason that some proxies produce few samples is because they are functional for only a few hours, so that we cannot have sufficient samples from them.

*selection frequency* for their $k$th most selected clusters as

$$F(k) = \frac{\sum_v f_v(k)}{N},$$
(4.1)

where $k = 1, 2, \cdots$, and $N$ is the total number of the VPs under study.

Figure 4.3 presents the normalized cluster selection frequencies. From the figure one can see that although multiple server clusters on the CDN network were selected for a VP over time, Youku did not distribute the workload among them evenly, but route most requests to only a few clusters. In addition, we can see that the selection frequency decreases rapidly as the rank increases (Fig. 4.4).

### 4.2.2 Understanding Server Selection Dynamics

We seek to understand the server selection dynamics exhibited by Youku through an experiment. In our experiment, we employ the methodology described in Sect. 4.1 to make frequent video requests to Youku from a measurement agent that is located at a fixed location and ISP network. We probed Youku in every two minutes, and collected the video server IP addresses that were returned in the 302 re-direction responses. The experiment lasted 24 h.

We have observed 15 distinct content server IP addresses in the 24-h measurement. These addresses are coming from five different clusters in five different cities.

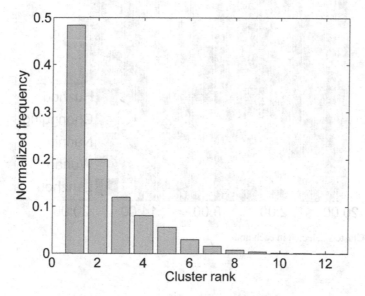

**Fig. 4.4** Normalized cluster selection frequencies across all the VPs

**Table 4.1** Ratios of the clusters being selected in 24 h

| Wuhan | Nanning | Chongqing | Huzhou | Lanzhou |
|-------|---------|-----------|--------|---------|
| 0.748 | 0.143 | 0.090 | 0.014 | 0.004 |

Table 4.1 presents the ratios of the clusters being selected, from which we can see that most requests were directed to the Wuhan cluster. In addition, we find that all the CDN clusters are in the same ISP of the measurement agent.

We then analyze the server selections on an hourly basis. Figure 4.5 presents the numbers of the times that the clusters were selected in each hour. From the figure we can see that Youku exhibited different degrees of dynamics in different hours: in the periods of 1:00–6:00 and 15:00–17:00, Youku consistently selected the most preferred Wuhan cluster; however, during the other hours, the CDN made dynamic decisions by switching among two or three clusters. Note that the hours with the greatest server selection dynamics (i.e., 8:00–13:00 and 17:00–23:00) are in fact the times that Youku attracts the most visits in a day, we can see that during these times, Youku has to balance the CDN workload among several clusters, which leads to the observed dynamics.

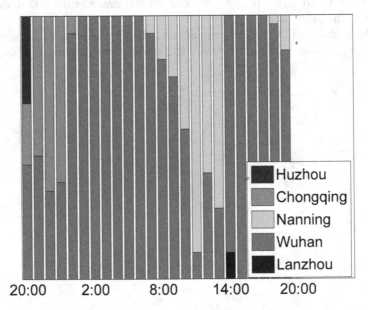

**Fig. 4.5** Clusters selected in each hour

### 4.2.3 ISP-Friendliness

For investigating the ISP-friendliness in Youku's server selections, we employ an indicator function $I(\cdot)$ for labeling the relationship between a VP and a CDN's server cluster: for a VP $v$, if its $k$th most selected cluster $c_k^v$ is in the same ISP of $v$, then $c_k^v$ is considered as a *home cluster* to $v$, and we have $I(v, c_k^v) = 1$; otherwise, $I(v, c_k^v) = 0$, and $c_k^v$ is referred to as a *foreign cluster*. With the indicator function, for all the VPs under study, we can define an aggregated *ISP-friendly ratio* for their $k$th most selected server clusters as

$$R(k) = \frac{\sum_v I(v, c_k^v) \cdot f_v(k)}{\sum_v f_v(k)} \tag{4.2}$$

Figure 4.6 presents the ISP-friendly ratios of the server clusters selected by Youku for all the 229 VPs, and we also present the ISP-unfriendly ratios as $1 - R(k)$. From the figure one can see that in most cases, the CDN follows an ISP-friendly policy by selecting home clusters. To better support this claim, we examine each VP's most preferred cluster: among the 229 VPs, only 20 of them select the foreign clusters as their most preferred clusters, and further examination shows that 15 of them are indeed in small stub ASes, in which Youku does not place any video servers. In other words, there are no home clusters on Youku's CDN for these VPs. An other observation from the figure is that the ISP-friendly ratio decreases as the cluster rank increases. Recall that Youku makes dynamic server selections for balancing the workloads during the busy hours, we can see that when a VP's home clusters are all

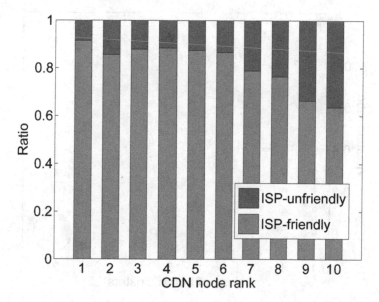

**Fig. 4.6** ISP-friendliness in Youku's server selections

overloaded, Youku will select the foreign clusters for accommodating the excessive workload and violate its ISP-friendly policy.

### 4.2.4   An ISP View of ISP-Friendliness Violation

Although it is observed that Youku follows an ISP-friendly policy in its server selections, however, in 6.2% of the samples, we find that the policy is indeed violated. Here we seek to study such a phenomenon from an ISP perspective. For each VP in our study, we define its *violation ratio* as the times that a foreign cluster was selected divided by the total number of the samples collected from the VP, and for each VP, we correlate its violation ratio with the number of its home clusters available for selection on the CDN network.

Figure 4.7 presents the correlation using a bubble graph, where each bubble on the figure represents a group of the VPs in a same ISP. For each bubble, x-axis indicates how many home clusters available for the VPs in the group, y-axis shows the violation ratio observed from these VPs, and the bubble size reflects the group size. Two observations could be made from the figure: First, for the ISPs with many server clusters, nearly no violation is observed. For example, 87 of the 94 VPs in the ISP of ChinaNet (corresponding to the big bubble at (27, 0)) and 70 of the 78 VPs in the ISP of China169 (the big bubble at (15, 0)) have a zero violation ratio. Second, for some small ISPs with few clusters available for selection, there are considerable violations. The observations suggest that Youku may have insufficient

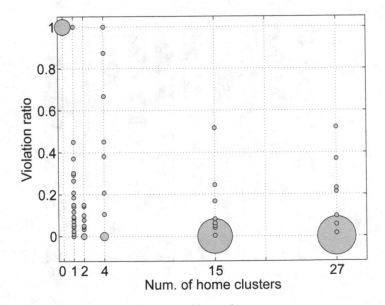

**Fig. 4.7** Correlation between violation ratio and home clusters

service capacities in these small ISPs, which forces the CDN to violate its ISP-friendly policy by selecting foreign clusters for the clients in these ISPs.

We summarize our observations as the following:

1. Youku employs the "deep-into-ISPs" design in its CDN network by deploying its server clusters in as many as 43 cities and 17 ASes in China;
2. Youku generally follows an ISP-friendly policy by preferring the server clusters that are in a same ISP for a client;
3. When handling the excessive workloads from the ISPs, the CDN violates the ISP-friendly policy, especially in the small ISPs in which Youku does not have sufficient service capacity.

### 4.2.5  Discussion

Through our measurement study, we have observed that Youku follows an ISP-friendly policy in selecting clients' server clusters. We also find that other large scale video streaming CDNs in China, like Tudou and Sina video, are employing similar policies. The reasons are in two-fold: on one hand, the ISPs have incentive to bring the CDN at "home" by hosting the CDN's server clusters at their data centers with relative lower price, as long as the CDN follows the ISP-friendly sever selection policy. On the other hand, when the CDN becomes less friendly to the ISPs, the ISPs can migrate the increased cross-ISP traffic cost to the CDN. For example, the ISP can directly charge higher prices on the CDN for renting servers and racks in their data centers; or the ISPs can throttle the CDN applications, which causes the CDN to have more failures in meeting the SLA requirement, and eventually lose its customers and profits. In either cases, CDN actually pays for their cross-ISP video deliveries. Therefore, it is very necessary for a CDN like Youku to select video servers in an ISP-friendly way for saving its operating cost.

## 4.3  Energy-Aware Capacity Provisioning

As a large scale Internet infrastructure, a large portion of CDN's operating cost is on energy. In recent years, many works [5–7] were focused on improving a CDN or a cloud's energy efficiency with techniques that dynamically "right-size" the system's service capacity. Generally, an energy-aware capacity provisioning technique is based on two facts: First, for many Internet services, the workload patterns are periodic, and exhibit great fluctuations, for example, the average workload for a large cloud service could be only 40% of the peak load; Second, current servers are far from energy-proportional, for instance, an idle server consumes up to 60% of the energy consumed by a full-loaded server [8]. Therefore, during the hours with lower workloads, by scheduling idle servers into the power-saving mode, considerable energy cost can be saved.

A typical energy-aware service capacity provisioning algorithm works as the following: Given a period of interest $t \in \{1, ..., T\}$, the mean workload at interval $t$ is denoted as $\mathbb{E}[x_t]$, the algorithm determines $n_t$, which is the service capacity provisioned during interval $t$, so that $n_t$ is larger than the actual workload $\mathbb{E}[x_t]$, and the overall energy cost is minimized. Sometimes, it is also desirable to reduce the *server switches*, that is, to reduce the times that servers are toggled into and out of the power-saving mode.

## 4.4  Implication and Motivation

From the above analysis, we can see that for a Youku-like CDN, or a CDN following the "deep-into-ISPs" design, there are two major components in its operating cost: (1) the traffic cost for cross-ISP content deliveries, and (2) the energy cost for running its server clusters. Note that *it is conflicting to reduce the both costs simultaneously*. To avoid the cross-ISP traffic, for each ISP, we need to reduce the chance that workload from the ISP exceeds the service capacity provisioned by the clusters in that ISP. This can be achieved by allocating some surplus capacity that is larger than the predicted workload. However, the power cost also rises with the over-provisioned capacity.

On the other hand, to reduce the energy cost, we need to right-size the active servers in each clusters based on the predicted workload. However, as workload fluctuates, in cases that an ISP's actual workload exceeds the planned service capacity, the CDN has to violate the ISP-friendly policy by selecting servers from foreign clusters, and incurs cross-ISP traffic cost. In the next chapter, we will present a solution that jointly saves the energy and the traffic costs for a video streaming CDN.

## References

1. Team cymru IP to ASN mapping. http://www.team-cymru.org/IP-ASN-mapping.html
2. Tian, Y., Dey, R., Liu, Y., Ross, K.W.: Topology mapping and geolocating for China's Internet. IEEE Trans. Parallel Distrib. Syst. **24**(9), 1908–1917 (2015)
3. Adhikari, V.K., Jain, S., Zhang, Z.-L.: YouTube traffic dynamics and its interplay with a tier-1 ISP: an ISP perspective. In: Proceedings of IMC, Melbourne, Australia, pp. 431–443 (2010)
4. Wang, Y.A., Huang, C., Li, J., Ross, K.W.: Estimating the performance of hypothetical cloud service deployments: a measurement-based approach. In: Proceedings of IEEE INFOCOM, Shanghai, China, pp. 2372–2380 (2011)
5. Lin, M., Wierman, A., Andrew, L.L.H., Thereska, E.: Dynamic right-sizing for power-proportional data centers. IEEE/ACM Trans. Netw. **21**(5), 1378–1391 (2013)
6. Mathew, V., Sitaraman, R.K., Shenoy, P.: Energy-aware load balancing in content delivery networks. In: Proceedings of IEEE INFOCOM, Orlando, FL, USA, pp. 954–962 (2012)
7. Tchernykh, A., Cortes-Mendoza, J., Pecero, J.E., Bouvry, P., Kliazovich, D.: Adaptive energy efficient distributed VoIP load balancing in federated cloud infrastructure. In: Proceedings of IEEE International Conference on Cloud Networking, Luxembourg, pp. 27–32 (2014)
8. Hameed, A., Khoshkbarforoushha, A., et al.: A survey and taxonomy on energy efficient resource allocation techniques for cloud computing systems. Computing **98**(7), 751–774 (2016)

# Chapter 5
# Capacity Provisioning for Video Content Delivery

From the previous chapter, we can see that for a video streaming CDN like Youku, there exists an inherent conflict between improving the CDN's energy efficiency and preserving its ISP-friendly server selection policy. In this chapter, we present a formal description of the problem, and propose a cost-aware capacity provisioning algorithm that can dynamically plan the CDN's service capacities in its clusters for saving the overall operating cost.

## 5.1 Problem Statement

### 5.1.1 The Network Model

We consider a Youku-like CDN that employs the "deep-into-ISPs" design. In particular, the CDN is composed of a number of server clusters, which are partitioned in $K$ different ISPs. An ISP can have multiple clusters, but for simplicity we assume that a CDN cluster can only be in one single ISP, that is, no multi-homing clusters. Each ISP has a number of point-of-presences (PoPs), where a PoP represents a group of clients that impose a large volume of aggregated workload upon the CDN.

The CDN allocates its capacity in time intervals. An interval could be in several minutes (for example, 10 min). In interval $t$, we denote the workload imposed on the CDN from all the PoPs in an ISP, say $ISP_i$, as $x_{i,t}$.[1] Note that $x_{i,t}$ can be viewed as a random variable, which fluctuates over time. At the beginning of each interval, the CDN plans the service capacities for its server clusters; more specifically, for $ISP_i$, the CDN determines $n_{i,t}$, the number of the servers that are scheduled to be alive in all its clusters in $ISP_i$, to provide the video streaming service.

---

[1] We list the notations used in describing the problem and our proposed algorithm in Table 5.1 in Appendix of this chapter.

© The Author(s) 2017
Y. Tian et al., *Internet Video Data Streaming*,
SpringerBriefs in Computer Science, DOI 10.1007/978-981-10-6523-1_5

The CDN under study follows an ISP-friendly server selection policy as observed in Sect. 4.2. That is, the CDN always selects a server from a home cluster for a client, as long as such a server is alive and has spare capacity. The CDN violates the policy only when all the servers in the client's home clusters are busy, and in that case, the CDN will deliver the video chunks from a server with spare capacity in a foreign cluster, and incurs cross-ISP traffic.

## 5.1.2   Cost Function

With the assistance of the network model, we formulate a CDN's overall operating cost. For one particular ISP, say $ISP_i$, it is easy to see that the energy cost of all its server clusters during interval $t$ is $c_1 \times n_{i,t}$, where $c_1$ is the power cost for running one active server per interval. For the CDN's cross-ISP traffic cost, note that cross-ISP video delivery happens only when the workload demand exceeds the service capacity in an ISP. By assuming one server providing one unit service capacity, the incurred cross-ISP traffic cost can be expressed as $c_2 \times \int_{n_{i,t}}^{\infty} x \times f_{i,t}(x)dx$, where $f_{i,t}(x)$ is the distribution of the workload $x_{i,t}$ during interval $t$, $\int_{n_{i,t}}^{\infty} x \times f_{i,t}(x)dx$ is the part of the workload that exceeds the service capacity in $ISP_i$, and $c_2$ is the cost for delivering one unit workload traffic in the cross-ISP way. Finally, the CDN's total cost for operating the clusters in $ISP_i$ during interval $t$ can be expressed as

$$c_{ISP}(n_{i,t}) = c_1 \times n_{i,t} + c_2 \times \int_{n_{i,t}}^{\infty} x \times f_{i,t}(x)dx \qquad (5.1)$$

From the above formulation, we can see that the service capacity in each ISP need to be carefully planned: by increasing $n_{i,t}$, the cross-ISP traffic cost will be reduced, but at a higher energy cost; while decreasing $n_{i,t}$ for energy-saving will lead to a higher cross-ISP traffic cost.

## 5.1.3   Problem Formulation

In our CDN capacity provisioning problem, we seek to plan and allocate the CDN's service capacity in each ISP, by determining $n_{i,t}$, to achieve the following objectives:

- **Meeting SLA**: The CDN should meet its Service Level Agreement (SLA). A typical SLA for a CDN service is its availability [1]. In this work, we consider SLA as the availability of the video service. That is, in any interval $t$, SLA requires that

$$\Pr[X_t < N_t] > th_{SLA}, \quad 0 < th_{SLA} \leq 1, \qquad (5.2)$$

where $X_t = \sum_{i=1}^{K} x_{i,t}$ is the overall workload from all the ISPs in interval $t$, $N_t = \sum_{i=1}^{K} n_{i,t}$ is the CDN's global service capacity, and $th_{SLA}$ is the SLA requirement in terms of the service availability.

- **Saving the operating cost**: The CDN should be cost effective, that is, its overall cost for operating the server clusters in all the ISPs, which can be expressed as

$$C(n_{1,t}, \cdots, n_{K,t}) = \sum_{i=1}^{K} c_{ISP}(n_{i,t}), \tag{5.3}$$

should be minimized. Note that $c_{ISP}(n_{i,t})$ contains both the energy cost and the cross-ISP traffic cost as indicated in Eq. (5.1).

Besides the two objectives, it is also expected that there are limited *server switches*, so as to reduce the wear-and-tear effects on the servers when switching them into and out of the power-saving mode.

## 5.2 Capacity Provisioning Algorithm

In this section, we propose our solution for the above described CDN capacity provisioning problem. The solution works in two steps: in the first step, we determine $\bar{N}_t$, the lower bound of the global service capacity, for meeting the CDN's SLA requirement; in the second step, we decide $n_{i,t}$, which is the number of the live servers in each ISP, for saving the CDN's overall operating cost. Obviously, $\sum_{i=1}^{K} n_{i,t} \geq \bar{N}_t$.

As in other capacity provisioning algorithms (e.g., [1–3]), our approach relies on predicting of the future workload. In particular, at the beginning of each interval, say interval $t$, the CDN scheduler predicts $f_{i,t}(x)$, the distribution of the workload from $ISP_i$, for each of the ISPs in which the CDN has deployed its clusters; and the scheduler also predicts the CDN's global workload distribution as $f_{G,t}(x)$.

We note that the prediction is feasible because of two reasons: First, recent studies show that by applying regression-based techniques and by employing sufficient history data, it is possible to accurately predict a video streaming service's average workload in a median-length interval like 10 min [3, 4]. Second, studies show that the instantaneous workload can be well approximated as modified Poisson [5] or Gaussian [6]; in Sect. 5.2.1, we will observe that Youku's instantaneous workload can be approximated as Gaussian, and similar to web traffic [7], we observe that the workload variance scales linearly with the mean workload. By combining these results, we can see that it is possible to predict the distribution of a CDN's global or cluster-wise instantaneous workloads with high accuracies.

### 5.2.1   Characteristics of CDN Workload

In this Section, we seek to capture the key characteristics of Youku's workload that can assist us to evaluate our proposed CDN service capacity provisioning scheme. More specifically, we testify if the workload models derived from other CDNs or web-based services can be applied on Youku.

Our study is based on a trace of the workload on Youku from a campus network. More specifically, we setup a traffic capturing program based on tcpdump at the gateway of a university network in China, and collected TCP in 24 h. From the trace we have found 256,183 requests for Youku video chunks. The workload is presented in Fig. 5.1. From the figure we can see that there is an obvious diurnal pattern, where the peak workload is over two times of the average. Furthermore, the workload is not smooth, but fluctuates dynamically, even within a short period of time.

Since the aggregated CDN workload can be viewed as generated from a large number of individual clients, one might expect its distribution to be close to Gaussian. Moreover, a recent study shows that CDN workload can be well approximated as Gaussian [6]. We use our campus workload trace to testify this argument, and find that it is indeed the truth. For example, in Fig. 5.2 we consider 5-s workload samples in two intervals, each last 30 minutes, and compare the empirical workload distributions with the Gaussian ones. We can see that the two distributions are very close. In fact, it is observed that samples over time slots of 10 and 30 s and in other time intervals are also nearly normally distributed.

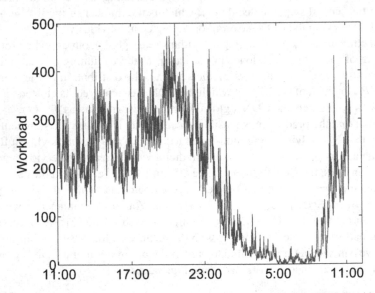

**Fig. 5.1** Workload from clients in a campus network imposed on Youku's CDN (in video chunk requests per minute)

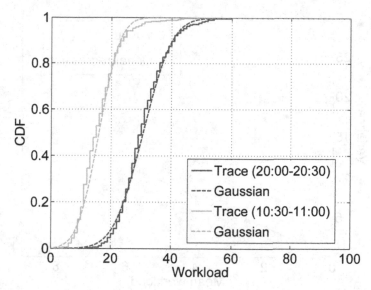

**Fig. 5.2** Distributions of 5-s workloads from 10:30–11:00 and 20:00–20:30 comparing with Gaussian distributions with the same means and standard deviations

We then investigate how the workload fluctuates. Note that for Web traffic, it is observed that the Gaussian traffic variance scales linear with the mean traffic [7]. We expect that such a scaling law also applies for the workload on Youku's CDN. To testify this, we consider a time slot of 1 s, and for each minute, we compute the mean and variance of the chunk requests in its 60 s. The correlation of the mean and variance of the 1-s workloads is presented in Fig. 5.3. From the figure, one can see that the workload variance scales nearly linearly with the mean workload. In fact, by using linear regression [8], the relationship between the mean workload and the workload variance can be approximated as

$$\mathrm{Var}(x) \approx a \cdot \mathbb{E}[x], \tag{5.4}$$

where $a \approx 2.21$ is a constant.

## 5.2.2  Estimating Global Capacity Lower Bound

As it is required that $\Pr[X_t < N_t] > th_{SLA}$ for meeting the CDN's SLA, with the prediction of the global workload distribution $f_{G,t}(x)$, it is easy to see that the lower bound of the global service capacity $\bar{N}_t$ can be obtained by solving the following problem

$$\int_0^{\bar{N}_t} f_{G,t}(x)dx = th_{SLA} \tag{5.5}$$

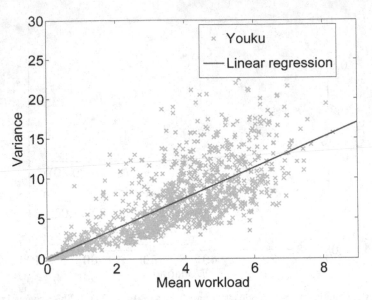

**Fig. 5.3** Correlation of 1-s mean workload (x-axis) and workload variance (y-axis)

### 5.2.3 Optimal ISP Capacity Allocation for Cost Saving

For minimizing the CDN's overall operating cost as in Eq. (5.3), we search for the optimal service capacities $\{n_{i,t}\}_{i=1}^{K}$ that are allocated in all the ISPs. The problem can be formulated as

$$\begin{aligned} & Minimize \ C(n_{1,t}, \cdots, n_{K,t}) = \sum_{i=1}^{K} c_{ISP}(n_{i,t}) \\ & s.t. \qquad \sum_{i=1}^{K} n_{i,t} \geq \bar{N}_t, \quad i = 1, \ldots, K \end{aligned} \qquad (5.6)$$

where $c_{ISP}(n_{i,t})$ is the CDN's operating cost in $ISP_i$ as expressed in Eq. (5.1), and $\bar{N}_t$ is the lower bound of the CDN's global capacity.

The problem is generally difficult to solve under arbitrary workload, however, for many well-known workload models such as exponential [5] and Gaussian [6], we find that $C(n_{1,t}, \cdots, n_{K,t})$ is convex, which makes the problem a convex optimization problem [9], whose global optimal solution can be obtained by greedily reducing the objective function until convergence. We use $\{n_{i,t}^{*}\}_{i=1}^{K}$ to denote the optimal solution for the convex optimization problem in Eq. (5.6).

### 5.2.4 Allocating ISP Capacities with Reduced Server Switches

The optimal solution for Eq. (5.6) can be used to determine the initial capacity of a CDN. However, during the regular operation, it is not applicable. This is because the algorithm is unaware of the server switches, therefore will frequently toggling servers into and out of the power-saving mode in consecutive intervals. To address this problem, we propose a heuristic algorithm that balances the cost saving with server switches.

After obtaining the global capacity lower bound $\bar{N}_t$, our proposed heuristic works in three phases iteratively to determine the capacity in each ISP. During each iteration, say iteration $j$ ($j = 0, 1, 2, \cdots$), the algorithm updates the temporarily allocated service capacity $n_{i,t}^{(j)}$ for $ISP_i$. Following is the algorithm details:

**Phase I: Initialization**: In the initial iteration, we let

$$n_{i,t}^{(0)} = \begin{cases} n_{i,t-1}, & \text{If } n_{i,t-1} \geq \lceil \mathbb{E}[x_{i,t}] \rceil \\ \lceil \mathbb{E}[x_{i,t}] \rceil, & \text{If } n_{i,t-1} < \lceil \mathbb{E}[x_{i,t}] \rceil \end{cases}, \tag{5.7}$$

where $i = 1, 2, ..., K$, and $\mathbb{E}[x_{i,t}]$ is the predicted mean workload in interval $t$. That is, the initial planned capacity should be no less than the predicted mean workload of the incoming interval, and can be the capacity in the previous interval when it is greater than $\mathbb{E}[x_{i,t}]$.

**Phase II: Saving CDN cost**: In the $j$th iteration ($j \geq 0$), for $ISP_i$, compare the current capacity $n_{i,t}^{(j)}$ with the optimal solution $n_{i,t}^*$ for the problem in Eq. (5.6):

- If $n_{i,t}^{(j)} < n_{i,t}^*$, and the cost reduction by adding one more server, $\Delta c_{i,t}^+ = c_{ISP}(n_{i,t}^{(j)}) - c_{ISP}(n_{i,t}^{(j)} + 1)$, is larger than a threshold $\theta$, we let $n_{i,t}^{(j+1)} = n_{i,t}^{(j)} + 1$ for $ISP_i$;
- Similarly, if $n_{i,t}^{(j)} > n_{i,t}^*$, and the cost reduction by removing one server, $\Delta c_{i,t}^- = c_{ISP}(n_{i,t}^{(j)}) - c_{ISP}(n_{i,t}^{(j)} - 1)$, is larger than the threshold $\theta$, we let $n_{i,t}^{(j+1)} = n_{i,t}^{(j)} - 1$.

For each ISP, repeat until no server can be added or removed any more.

**Phase III: Meeting SLA**: In this phase, compare the current global capacity $\sum_{i=1}^{K} n_{i,t}^{(j)}$ with the global capacity lower bound $\bar{N}_t$, if $\sum_{i=1}^{K} n_{i,t}^{(j)} < \bar{N}_t$, which means more servers should be alive for meeting the SLA requirement, the heuristic finds the ISP that has the minimum cost increase (or the maximum cost reduction) by adding one more server, that is, find

$$s = \arg \max_{i=1,...,K} \{\Delta c_{i,t}^+\} \tag{5.8}$$

and add one more server for $ISP_s$ by letting $n_{s,t}^{(j+1)} = n_{s,t}^{(j)} + 1$; repeat until $\sum_{i=1}^{K} n_{i,t}^{(j)} \geq \bar{N}_t$.

Note that in Phase II, we compare the expected cost reduction by adding or removing one server with threshold $\theta$, and actually add or remove a server only when it is worthwhile. In fact, we can view $\theta$ as the wear-and-tear cost of a server for making a server switch, and the algorithm avoids unnecessary switches by hibernating or awakening a server only when the benefit is significant enough. A formal description of the algorithm can be found in Algorithm 1.

---

**Algorithm 1** CDN capacity provisioning algorithm

1: $n_{i,t}^{(0)} \leftarrow \begin{cases} n_{i,t-1}, & n_{i,t-1} \geq \lceil \mathbb{E}[x_{i,t}] \rceil \\ \lceil \mathbb{E}[x_{i,t}] \rceil, & n_{i,t-1} < \lceil \mathbb{E}[x_{i,t}] \rceil \end{cases}$;     ▷ Phase I

2: $j \leftarrow 0$;

3: **repeat**     ▷ Phase II

4:     **for** $i \leftarrow 1, K$ **do**

5:

6:         **if** $n_{i,t}^{(j)} < n_{i,t}^*$ & $\Delta c_{i,t}^+ \geq \theta$ **then**

7:             $n_{i,t}^{(j+1)} \leftarrow n_{i,t}^{(j)} + 1$;

8:         **else**

9:             **if** $n_{i,t}^{(j)} > n_{i,t}^*$ & $\Delta c_{i,t}^- \geq \theta$ **then**

10:                 $n_{i,t}^{(j+1)} \leftarrow n_{i,t}^{(j)} - 1$;

11:             **end if**

12:         **end if**

13:     **end for**

14:     $j \leftarrow j + 1$;

15: **until** no server can be added or removed

16: **while** $\sum_{i=1}^{K} n_{i,t}^{(j)} < \bar{N}_t$ **do**     ▷ Phase III

17:     $s \leftarrow \arg\max_{i=1,\dots,K} \{\Delta c_{i,t}^+\}$;

18:     $n_{s,t}^{(j+1)} \leftarrow n_{s,t}^{(j)} + 1$; $j \leftarrow j + 1$;

19: **end while**

20: $n_{i,t} \leftarrow n_{i,t}^{(j)}$;

---

# Appendix

The notations used in describing the problem and the proposed algorithm are listed in Table 5.1.

**Table 5.1** Notations used in problem and algorithm description

| Notation | Meaning |
|---|---|
| $x_{i,t}$ | CDN workload from $ISP_i$ in interval $t$; |
| $f_{i,t}(x)$ | Distribution of the workload of $x_{i,t}$; |
| $n_{i,t}$ | Number of active servers planned in $ISP_i$ in interval $t$; |
| $c_{ISP}(n_{i,t})$ | CDN's operating cost in $ISP_i$ in interval $t$ as defined in Eq. (5.1); |
| $th_{SLA}$ | SLA in terms of the video service availability; |
| $X_t$ | CDN workload from all the ISPs in interval $t$; |
| $f_{G,t}(x)$ | Distribution of the workload of $X_t$; |
| $N_t$ | Total number of active servers planned in interval $t$; |
| $C(n_{1,t}, \cdots)$ | CDN's operating cost in all the ISPs in interval $t$ as defined in Eq. (5.3); |
| $\bar{N}_t$ | Lower bound of the global service capacity for meeting SLA in interval $t$; |
| $n_{i,t}^*$ | Number of active servers for $ISP_i$ in interval $t$ by solving Eq. (5.6); |
| $n_{i,t}^{(j)}$ | Temporarily allocated service capacity for $ISP_i$ during the $j^{th}$ iteration in Algorithm 1; |

# References

1. Mathew, V., Sitaraman, R.K., Shenoy, P.: Energy-aware load balancing in content delivery networks. In: Proceedings of IEEE INFOCOM, Orlando, FL, USA, pp. 954–962 (2012)
2. Lin, M., Wierman, A., Andrew, L.L.H., Thereska, E.: Dynamic right-sizing for power-proportional data centers. In: Proceedings of IEEE INFOCOM, Shanghai, China, pp. 1098–1106 (2011)
3. Niu, D., Feng, C., Li, B.: Pricing cloud bandwidth reservations under demand uncertainty. In: Proceedings of ACM SIGMETRICS, London, UK, pp. 151–162 (2012)
4. Niu, D., Liu, Z., Li, B., Zhao, S.: Demand forecast and performance prediction in peer-assisted on-demand streaming systems. In: Proceedings of IEEE INFOCOM, Shanghai, China, pp. 421–425 (2011)
5. Kang, X., Zhang, H., Jiang, G., Chen, H., Meng, X., Yoshihira, K.: Understanding Internet video sharing site workload: A view from data center design. J. Vis. Commun. Image Rep. **21**(2), 129–138 (2010)
6. Bak, A., Gajowniczek, P., Pilarski, M.: Gaussian approximation of cdn call level traffic. In: Proceedings of the International Teletraffic Congress, San Francisco, CA, USA, pp. 135–141 (2011)
7. Morris, R., Lin, D.: Variance of aggregated web traffic. In: Proceedings of IEEE INFOCOM, TelAviv, Israel, pp. 360–366 (2000)
8. Murphy, K.P.: Machine Learning: A Probabilistic Perspective. The MIT Press, Cambridge, MA, USA (2012)
9. Boyd, S., Vandenberghe, L.: Convex Optimization. Cambridge University Press, Cambridge, UK (2004)

# Chapter 6
# Performance Evaluation

In this chapter, we evaluate our proposed CDN capacity provisioning algorithm, and compare it with other schemes through simulation experiments. We use the workload derived from real-world measurement, the actual bandwidth and power prices in our simulated experiment.

## 6.1 Experiment Setup

We emulate a CDN that deploys its video server clusters in 14 different ISPs. ISPs vary in size, and we consult the measurement result in Sect. 4.2 by assigning various numbers of PoPs in different ISPs, where the largest ISP has 88 PoPs and the smallest has only one PoP. Each PoP imposes a certain amount of video requests on the CDN per second, and the CDN follows the ISP-friendly policy by directing a request to an active server in the same ISP as much as possible. However, when all the servers are overloaded, the request will be handled by a server from a different ISP, and incurs some cross-ISP traffic cost.

The CDN plans its service capacity in intervals. An interval lasts for 10 min in our simulation. At the beginning of an interval, the CDN predicts the workload of the incoming interval, and applies one of the following capacity provisioning schemes to decide the number of the active servers for each ISP:

- The *energy-aware* capacity provisioning: In this approach, during each interval, after obtaining the lower bound $\bar{N}_t$ of the global service capacity, the CDN allocates a capacity for each ISP that is proportional to the predicted workload from the ISP, that is, for $ISP_i$, $n_{i,t} = \lceil \frac{E[x_{i,t}]}{E[X_t]} \times \bar{N}_t \rceil$. Since only the lower bound capacity is provisioned, the CDN's energy expense is minimized. Note that such a scheme consumes even less power than existing energy-aware solutions (e.g., [1, 2]) as we do not consider reducing the server switches here.

© The Author(s) 2017
Y. Tian et al., *Internet Video Data Streaming*,
SpringerBriefs in Computer Science, DOI 10.1007/978-981-10-6523-1_6

- The *optimal* capacity provisioning: In this approach, the CDN allocates the service capacities for the ISPs by solving the convex optimization problem in Eq. (5.6) in each interval.
- The heuristic *cost-aware* capacity provisioning: In this scheme, we use the optimal solution of Eq. (5.6) for the initial interval, then apply the heuristic algorithm described in Sect. 5.2.4 to plan the CDN capacity in each ISP in the subsequent intervals. Note that the heuristic employs a threshold $\theta$. Since it is the wear-and-tear cost of a server switch, $\theta$ can be expressed as $\rho$ times of the unit chunk energy cost by letting $\theta = \rho \times c_1$, we consider various values of the parameter $\rho$ in our simulation.

For the CDNs employing different capacity provisioning schemes, we examine and compare the following performance metrics from one-day operating of the CDN:

- The CDN's monetary *operating cost*, including both the energy cost as well as the cross-ISP traffic cost. We will explain how to compute the two costs later in this subsection.
- The service capacity provisioned by the CDN. In particular, we are interested in the *over-provision ratio* of the CDN's capacity, which is defined as the ratio between the service capacity determined by the algorithm under study divided by the minimum capacity $\bar{N}_t$ for saving the energy cost only. In other words, the over-provision ratio indicates how many extra servers the CDN should keep alive for the objectives of saving its overall operating cost and reducing the server switches.
- The *mean-time-between-switch* (MTBS) for the servers in the CDN, which is the mean time between two consecutive switches of a video server.

From the definitions, we can see that a preferred capacity provisioning scheme should be able to reduce the CDN's operating cost, but it should also avoid frequent server switches by achieving a medium-long MTBS.

We use synthetic workload traces derived from the real-world workload on Youku that we have collected from a campus network in Sect. 5.2.1 to feed our simulator. More specifically, we view each PoP as a mixture of 20 independent workload sources. Each source imposes a workload whose mean is randomly drawn from the empirical distribution of the campus trace in recent 30 min, and the workload fluctuates following a Gaussian distribution whose variance scales linearly with the mean workload, as shown in Sect. 5.2.1.

We assume that videos are requested and delivered in 256 KB-sized chunks, and each video server has a maximum data rate of 100 Mbps. To compute the energy cost, we adopt the energy model in [2], which states that a server's energy consumption is $(63 + 29 \times U)$ watts, where $U$ is the server's utilization ratio. We then refer the current power price for data centers in China and compute that $c_1$, the energy cost for serving one video chunk, is about $9.14 \times 10^{(-7)}$ US dollars.

To compute the cross-ISP traffic cost, we need to decide the value of $c_2$, the cross-ISP traffic cost for delivering one video chunk. In fact, $c_2$ depends on the bandwidth price negotiated between the ISPs, and varies from a few to tens of the times of $c_1$. For simplicity, in our simulation we always let $c_2 = 5 \times c_1$ if not otherwise specified.

Finally, we require an availability of $th_{SLA} = 0.97$ as the CDN's SLA, since such an availability is typical in real-world video streaming services [3].

## 6.2 Evaluation and Comparison

### 6.2.1 Overall Performance

In our first experiment, we evaluate and compare the energy-aware, optimal, and cost-aware schemes. For the cost-aware scheme, we vary $\rho$ from 0.1 to 1.0. Figures 6.1, 6.2 and 6.3 presents the CDN's overall operating costs, the energy costs, and the cross-ISP traffic costs, under the three kinds of CDN capacity provisioning schemes respectively. Figure 6.4 shows the capacity over-provision ratios and Fig. 6.5 gives the MTBS of the servers under the different schemes.

From Figs. 6.1, 6.2 and 6.3, one can see that by jointly optimizing the energy cost and cross-ISP traffic cost, the optimal scheme saves 21.5% of the CDN's operating expense comparing with the energy-aware scheme. We also find that with moderate $\rho$ values, the cost-aware scheme has a cost between the energy-aware and the optimal schemes, and achieves 8.5–17.2% of the cost savings comparing with the scheme that focuses only on energy.

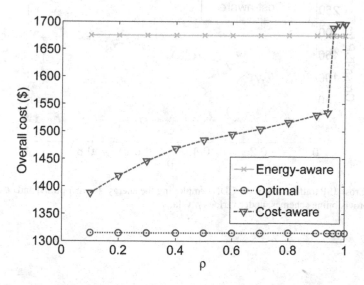

**Fig. 6.1** Overall costs of the CDNs employing the energy-aware, optimal and cost-aware capacity provisioning schemes, under various $\rho$ values

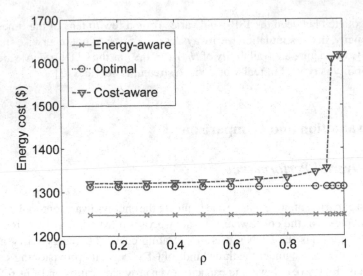

**Fig. 6.2** Energy costs of the CDNs employing the energy-aware, optimal and cost-aware capacity provisioning schemes, under various $\rho$ values

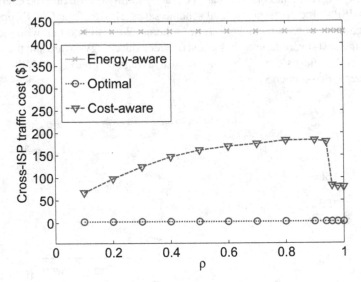

**Fig. 6.3** Cross-ISP traffic costs of the CDNs employing the energy-aware, optimal and cost-aware capacity provisioning schemes, under various $\rho$ values

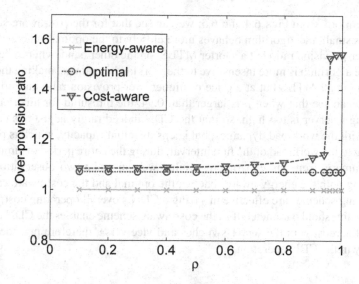

**Fig. 6.4** Over-provision ratios of the CDNs employing the energy-aware, optimal and cost-aware capacity provisioning schemes, under various $\rho$ values

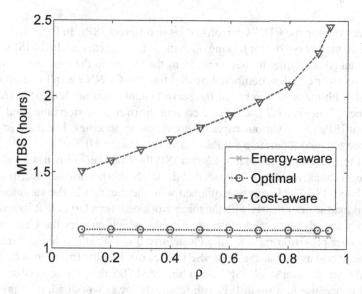

**Fig. 6.5** MTBS of the CDNs employing the energy-aware, optimal and cost-aware capacity provisioning schemes, under various $\rho$ values

Furthermore, from Figs. 6.4 and 6.5, we can see that for the cost-aware scheme, when $\rho$ is small, the algorithm behaves more closely to the optimal scheme, with a lower over-provision ratio but a shorter MTBS; on the other hand, when $\rho$ becomes larger, the algorithm is more insensitive to the workload dynamics, making the CDN to have a longer MTBS but at a price of higher over-provision ratio and operating cost. One can see that when $\rho$ is larger than 0.94, the threshold for hibernating or awakening a server is too high, so that the CDN indeed rarely adjusts its capacity to cope with the workload dynamics, but keeps its initial capacity, which is planned according to the workload of the first interval, during the entire period of simulation.

In summary, from the experiment results, we can make two observations: (1) Comparing with the energy-aware scheme, the optimal and the cost-aware capacity provisioning schemes are effective in saving a CDN's overall operating cost; (2) By tuning the threshold parameter of $\rho$, the cost-aware scheme enables the CDN to trade its operating cost with the server switches and vice versa, therefore provides more flexibility in the CDN operation.

### 6.2.2  Performance from ISP Perspective

We further examine the CDN's performances in different ISPs. In Fig. 6.6, we show the averaged monetary cost for serving one video chunk in each of the 14 ISPs, where ISPs are ranked according to their sizes from the largest to the smallest.[1] An ISP's averaged cost per chunk is computed by dividing the CDN's overall operating cost in this ISP with the total number of the served chunks that are requested from the PoPs in the ISP. In Figs. 6.7 and 6.8, we present the over-provision ratios and MTBS in different ISPs under various capacity provisioning schemes. For the cost-aware scheme, we choose moderate $\rho$ values as $\rho = 0.1, 0.4$, and 0.7.

From Fig. 6.6 one can see that in a larger ISP, the CDN generally has a lower cost per serving a chunk, and Fig. 6.7 shows that the CDN also has smaller over-provision ratios in larger ISPs. This can be explained with the fact that as the variance of the CDN workload scales linearly with the mean workload, for a larger ISP, its workload is relatively less dynamic and more predictable, which enables the CDN to have a smaller over-provision ratio for the clusters in this ISP, and achieve a relatively lower energy cost as well as the cross-ISP traffic cost. Finally, from Fig. 6.8, one can see that the servers in a small ISP have a longer MTBS than the servers in a larger ISP. This is because in a small ISP with relatively lower workload, in many cases, the absolute values of the workload changes are not large enough for hibernating or awakening one video server. Our observations here suggest that for a CDN employing the "deep-into-ISPs" design, it is more cost-effective to deploy the server clusters in larger ISPs than in smaller ones.

---

[1] We do not plot the energy-aware scheme for the reason that its costs are much larger than the ones of the other solutions.

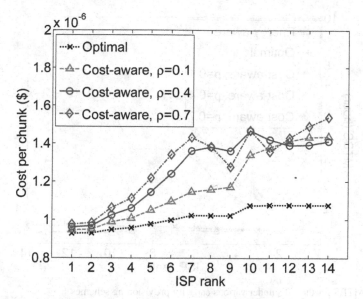

**Fig. 6.6** Costs per chunk for the ISPs under various capacity provisioning schemes

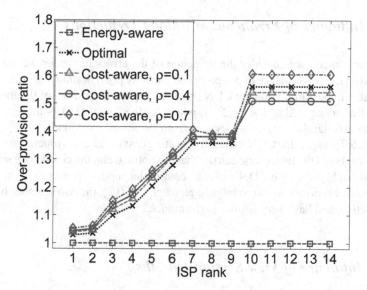

**Fig. 6.7** Over-provision ratios for the ISPs under various capacity provisioning schemes

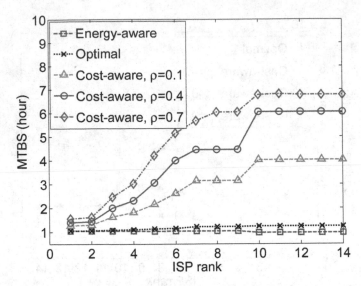

**Fig. 6.8** MTBS for the ISPs under various capacity provisioning schemes

## 6.2.3   Influence of Erroneous Workload Predictions

In this experiment, we consider the influence of the errors in predicting the future workload on the performance of our proposed capacity provisioning algorithm. More specifically, in our simulation the CDN predicts the mean workload of the incoming interval $t$ as $w_{i,t} = \mathbb{E}[x_{i,t}] \times (1 \pm e)$, where $e$ ($0 \leq e < 0.5$) is the error ratio indicating how far the prediction deviates from the actual workload.

Figure 6.9 presents the CDN's overall operating cost under the erroneous workload predictions. From the figure, one can see that the optimal and the cost-aware schemes can considerably save the CDN's overall cost under modest error ratios; however, when the error ratio becomes too large (e.g., when $e \geq 0.2$), the two schemes become less effective, and have very similar performance.

## 6.2.4   Influence of View Session Lengths

In our previous experiments, we assume that user requests video by chunks. In this experiment, we consider a scenario that a user only requests the video once, and the content streams to the client for a continuous period of time (which is defined as the user's *view session time*). A user's session time depends on many factors, such as the length of the video file, the level of interest held by the user on the content, etc., thus for each request, we can simply suppose its associated session length as random variable, which follows an exponential distribution. We vary users' mean session

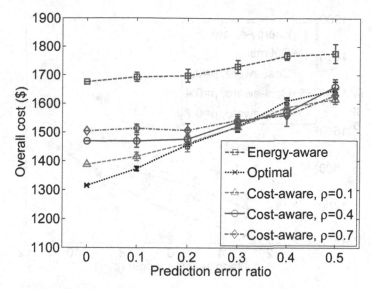

**Fig. 6.9** CDN's overall operating costs under erroneous workload predictions

length, and investigate its influence on CDN's operating cost. However, for different mean lengths, we tune the video request rates so that the total workload imposed on CDN is constant.

We plot the CDN's overall operating costs under the varying mean session lengths in Fig. 6.10. From the figure, one can observe that the CDN's operating cost rises as users tend to view the videos longer, and the cost-aware schemes constantly outperform the energy-aware scheme with lower costs; it is also interesting to see that the cost for the optimal scheme grows faster than the cost-aware schemes, and will surpasses the latter when the mean session length is longer than 10 min.

We seek to explain the observations as the following: Since a user views a video for a random period of time, the workload imposed by one video request on the CDN is also random, therefore for the CDN capacity provision schemes, it is difficult to have an accurate workload prediction under varying view session lengths, and for the exponential distribution, the longer the mean session length is, the less predictable the workload will be. For this reason, we can see that when users tend to have longer session lengths, the CDN pays more for the cross-ISP traffics due to the increasingly inaccurate workload predictions; moreover, such inaccuracy influences the optimal scheme more seriously than the cost-aware approaches, as the latter uses a threshold to avoid some unnecessary server switches.

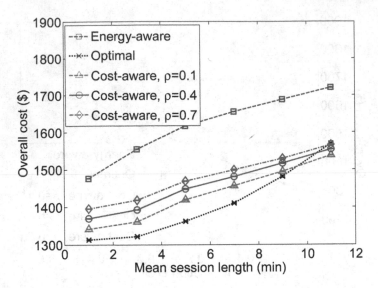

**Fig. 6.10** CDN's overall operating costs under varying video session lengths

### 6.2.5   Influence of Power and Bandwidth Prices

In the previous experiments we assume that by delivering a video chunk in a cross-ISP way, the incurred traffic cost is five times of the energy cost for serving the chunk, that is, $c_2 = 5 \times c_1$. In this experiment, we examine how the trends of the bandwidth and power prices influence the performances of the CDN.

In our experiment we suppose that the cross-ISP chunk delivery cost ($c_2$) varies from 4 to 8 times of the energy cost per serving a chunk ($c_1$). Figure 6.11 presents the CDN's over-provision ratios under the energy-aware, optimal and cost-aware schemes. From the figure one can see that when the cross-ISP bandwidth price is relatively higher than the power price, the CDN has to over-provide more capacities to avoid the cross-ISP content deliveries and save its overall operating cost.

We discuss the implication of Fig. 6.11 as the following: Although in the long-term future, it is expected that the energy price will continue to rise and the cross-ISP bandwidth price will continue to fall [4], however, under the current power and bandwidth prices, $c_2$ is still much higher than $c_1$, indicating that for a CDN with the "deep-into-ISPs" design, a significant part of its service capacity should be over-provisioned for avoiding the cross-ISP video deliveries. The observation also suggests that the current Internet business relationship among the ISPs [5] actually discourages a CDN to improve its energy-efficiency, and a new business model that is more energy-efficient and friendly to the environment should be negotiated between the ISPs and the CDNs.

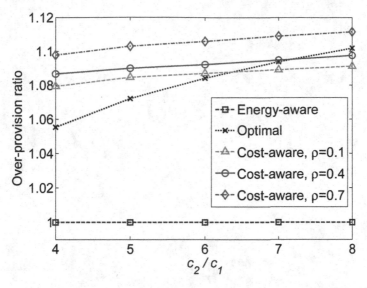

**Fig. 6.11** Over-provision ratios under different power and cross-ISP bandwidth prices

# References

1. Lin, M., Wierman, A., Andrew, L.L.H., Thereska, E.: Dynamic right-sizing for power-proportional data centers. In: Proceedings of IEEE INFOCOM, Shanghai, China, pp. 1098–1106 (2011)
2. Mathew, V., Sitaraman, R.K., Shenoy, P.: Energy-aware load balancing in content delivery networks. In: Proceedings of IEEE INFOCOM, Orlando, FL, USA, pp. 954–962 (2012)
3. Liu, X., Dobrian, F., Milner, H., Jiang, J., Sekar, V., Stoica, I., Zhang, H.: A case for a coordinated internet video control plane. In: Proceedings of ACM SIGCOMM, Helsinki, Finland, pp. 359–370 (2012)
4. Palasamudram, D.S., Sitaraman, R.K., Urgaonkar, B., Urgaonkar, R.: Using batteries to reduce the power costs of internet-scale distributed networks. In: Proceedings of ACM Symposium on Cloud Computing, San Jose, CA, USA, pp. 1–14 (2012)
5. Gao, L.: On inferring autonomous system relationships in the internet. ACM/IEEE Trans. Netw. **9**(6), 733–745 (2001)

References

# Chapter 7
# Concluding Remarks

We conclude the whole book in this chapter. The future research directions are also discussed.

## 7.1 Conclusions

With the technological advances on communications, multimedia, storage, and mobile Internet, more and more videos are distributed by Content Delivery Networks (CDNs) over the Internet, and how to reduce the operating cost of a video streaming CDN becomes an important issue. In this book, we survey large-scale CDNs in real world, and discuss their key issues and design choices. We also analyze the representative energy-saving techniques for server clusters, data centers, and CDNs. We focus on the Internet video streaming CDNs that employ a "deep-into-ISPs" design, and address the problem of saving a CDN's overall operating cost. By studying the CDN infrastructure of the largest Internet video site in China, namely Youku, we find that the CDN employs an ISP-friendly policy in selecting servers for users, and there exists an inherent conflict between improving the CDN's energy efficiency and maintaining its ISP-friendliness. Motivated by the observation, we propose a practical solution that seeks to save both the energy cost and the cross-ISP traffic cost for a CDN. Experiment results show that our approach can significantly reduce a CDN's overall operating cost, and enable the system to avoid frequent server switches effectively.

## 7.2 Future Research Directions

Although delivering majority of contents on the Internet, however, CDN is an Over-the-Top (OTT) solution, which has its inherited limitations. Given the prevalence of contents, especially videos in the future Internet, people from industry and

© The Author(s) 2017
Y. Tian et al., *Internet Video Data Streaming*,
SpringerBriefs in Computer Science, DOI 10.1007/978-981-10-6523-1_7

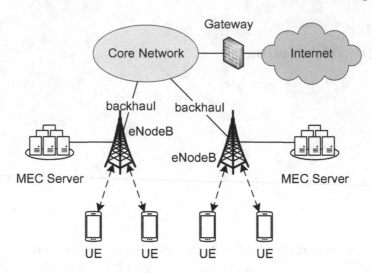

**Fig. 7.1**  Demonstration of MEC-enabled cellular network

academia propose that the content delivery functionality should be integrated in the future network's infrastructure. Among these proposals, we believe the following two topics are worth further investigation.

**Energy-aware caching for Mobile Edge Computing (MEC).** With the tremendous of wireless video traffic, backhaul network is becoming a bottleneck for mobile cellular systems. In the next generation broadband mobile wireless network (i.e., 5G), Mobile Edge Computing (MEC) is introduced to distribute cloud and storage capabilities, which used to be provided by Internet data centers, to the edge of the cellular radio access network (RAN). In particular, by deploying content servers directly at the base stations (BSs), higher user satisfaction can be meet and considerable workload can be offloaded from backhaul [1, 2]. A cellular network with MEC is demonstrated in Fig. 7.1.

For saving energy cost of cellular networks, most efforts seek to reduce the BS power, as BSs consume nearly 60–80% of the total RAN energy. Existing works focus on turning underutilized BSs into power-saving sleep mode, or adapting its transmit power to match traffic load. For example, Wu et al. [3] study the sleeping control and power matching (PM) for a single cell in cellular networks with bursty traffic; Yu et al. [4] investigate the energy saving problem of BSs in cellular networks, and propose a two-step solution. However, when MEC servers, in particular, content servers are co-located with BS, a cell's service capacity and its energy consumption are depending not only on the BS, but also on the MEC servers, and the new capacity-energy tradeoff requires further investigation.

**Green Information-centric Networking (ICN).** With content becoming the first class citizen, innovative Information-centric Networking (ICN) architectures have been proposed [5, 6]. Unlike the current Internet, which has the IP protocol as a thin

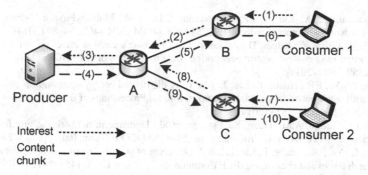

**Fig. 7.2** Demonstration of ICN routing and caching

waist of its protocol architecture, the content naming layer is the thin waist of ICN, and each content on the Internet has its unique name. The ICN nodes (clients, servers, and ICN routers) request, lookup, and route name requests and data using content names instead of the equipment addresses (e.g., IP addresses). For example, Fig. 7.2 demonstrates an example of ICN routing and caching based on content name.

Recently, a number of approaches are proposed for improving the ICN energy efficiency. For example, Choi et al. [7] consider to improve the ICN router energy cost by placing content on various cache hardwares. Fang et al. [8] propose to solve a Mixed Integer Linear Programming problem to switch off a subset of routers and links for energy saving purpose. Llorca et al. [9] present a solution to drive contents towards interested users along the minimum energy configuration, by guiding local caching decisions based on the global energy benefit. However, ICN is still in its early stage, and there are several challenges remain unresolved, such as the trade-off between energy saving and network performance, trade-off between energy saving and mobility, handling bursty flash crowds, etc. [10]. In addition, the feasibility of green ICN techniques needs to be testified in real world.

# References

1. Zeydan, E., Bastug, E., Bennis, M., et al.: Big data caching for networking: moving from cloud to edge. IEEE Commun. Mag. **54**(9), 36–42 (2016)
2. Peng, X., Zhang, J., Song, S.H., Letaief K.B.: Cache size allocation in backhaul limited wireless networks. In: Proceedings of IEEE ICC, Kuala Lumpur, Malaysia, pp. 1–6 (2016)
3. Wu, J., Bao, Y., Miao, G., Zhou, S., Niu, Z.: Base-station sleeping control and power matching for energy-delay tradeoffs with bursty traffic. IEEE Trans. Veh. Technol. **65**(5), 3657–3675 (2016)
4. Yu, N., Miao, Y., Mu, L., Du, H., Huang, H., Jia, X.: Minimizing energy cost by dynamic switching ON/OFF base stations in cellular networks. IEEE Trans. Wireless Commun. **15**(11), 7457–7469 (2016)
5. Zhang, L., Afanasyev, A., Burke, J., Jacobson, V., et al.: Named data networking. SIGCOMM Comput. Commun. Rev. **44**(5), 66–73 (2014)

6. Venkataramani, A., Kurose, J.F., Raychaudhuri, D., et al.: MobilityFirst: a Mobility-centric and trustworthy Internet architecture. ACM SIGCOMM CCR **44**(3), 74–80 (2014)
7. Choi, N., Guan, K., Kilper, D., Atkinson, G.: In-network caching effect on optimal energy consumption in content-centric networking. In: Proceedings of IEEE ICC, Ottawa, Canada, pp. 2889–2894 (2014)
8. Fang, C., Yu, F.R., Huang T., Liu, J., Liu Y.: A distributed energy optimization algorithm in content-centric networks via dual decomposition. In: Proceedings of IEEE Globecom, Austin, TX, USA, pp. 1848–1853 (2014)
9. Llorca J., Tulino, A.M., Guan, K., Esteban, et al.: Dynamic in-network caching for energy efficient content delivery. In Proceedings of IEEE INFOCOM, Turin, Italy, pp. 245–249 (2013)
10. Fang, C., Yu, F.R., Huang, T., Liu, J., Liu, Y.: A survey of green information-centric networking: research issues and challenges. IEEE Commun. Surv. Tutor. **17**(3), 1455–1472 (2015)

Printed in the United States
By Bookmasters